T0222433

A Second Course in Probability
Second Edition

Written by Sheldon M. Ross and Erol A. Peköz,this text familiarizes readers with advanced topics in probability while keeping the mathematical prerequisites to a minimum. Topics covered include measure theory, limit theorems, bounding probabilities and expectations, coupling and Stein's method, martingales, Markov chains, renewal theory, and Brownian motion. No other text covers all of these topics rigorously but at such an accessible level; only an undergraduate-level understanding of calculus and probability is needed.

New to this edition are sections on the gambler's ruin problem, Stein's method as applied to exponential approximations, and applications of the martingale stopping theorem. Extra end-of-chapter exercises have also been added, with selected solutions available. This is an ideal textbook for students taking an advanced undergraduate or graduate course in probability. It is also a useful resource for professionals in relevant application domains, from finance to machine learning.

SHELDON M. ROSS is the Epstein Chair Professor in the Epstein Department of Industrial and Systems Engineering at the University of Southern California. He has published more than 150 technical articles as well as a variety of textbooks in the areas of applied probability, statistics, and industrial engineering. He is the founding and continuing editor of the journal *Probability in the Engineering and Informational Sciences*, a fellow of the Institute of Mathematical Statistics and of the Institute for Operations Research and the Management Sciences, and a recipient of the Humboldt US Senior Scientist Award. He is the recipient of the 2006 INFORMS Expository Writing Award.

EROL A. PEKÖZ is Professor and Department Chair of Operations and Technology Management in the Questrom School of Business at Boston University. He has published more than 50 technical articles in applied probability and statistics and is the author of *The Manager's Guide to Statistics* (2009). In 2001, he was awarded Boston University's Broderick Prize for Teaching.

A Second Course in Probability

Second Edition

SHELDON M. ROSS
University of Southern California

EROL A. PEKÖZ
Boston University

CAMBRIDGE
UNIVERSITY PRESS

CAMBRIDGE
UNIVERSITY PRESS

Shaftesbury Road, Cambridge CB2 8EA, United Kingdom

One Liberty Plaza, 20th Floor, New York, NY 10006, USA

477 Williamstown Road, Port Melbourne, VIC 3207, Australia

314–321, 3rd Floor, Plot 3, Splendor Forum, Jasola District Centre,
New Delhi – 110025, India

103 Penang Road, #05–06/07, Visioncrest Commercial, Singapore 238467

Cambridge University Press is part of Cambridge University Press & Assessment,
a department of the University of Cambridge.

We share the University's mission to contribute to society through the pursuit of
education, learning and research at the highest international levels of excellence.

www.cambridge.org
Information on this title: www.cambridge.org/9781009179911

DOI: 10.1017/9781009179928

First and Second editions © Sheldon M. Ross and Erol A. Peköz 2007, 2023

First published 2007
Second Edition 2023 (Version 2, July 2023)

First printed in the United States of America 2007

A catalogue record for this publication is available from the British Library.

*A Cataloging-in-Publication data record for this book is available from the Library
of Congress*

ISBN 978-1-009-17991-1 Paperback

Contents

Preface

This book is intended to be a second course in probability for undergraduate and graduate students in statistics, mathematics, engineering, finance, and actuarial science. It is a guided tour aimed at instructors who want to give their students a familiarity with some advanced topics in probability, without having to wade through the exhaustive coverage contained in the classic advanced probability theory books (books by Billingsley, Chung, Durrett, Breiman, etc.). The topics covered here include measure theory, limit theorems, bounding probabilities and expectations, coupling, Stein's method, martingales, Markov chains, renewal theory, and Brownian motion.

One noteworthy feature is that this text covers these advanced topics rigorously but without the need for much background in real analysis; other than calculus and material from a first undergraduate course in probability (at the level of *A First Course in Probability*, by Sheldon Ross [7]), any other concepts required, such as the definition of convergence, the Lebesgue integral, and countable and uncountable sets, are introduced as needed.

The treatment is highly selective, and one focus is on giving alternative or nonstandard approaches for familiar topics to improve intuition. For example, we introduce measure theory with an example of a nonmeasurable set, prove the law of large numbers using the ergodic theorem in the first chapter, and later give two alternative (but beautiful) proofs of the central limit theorem using Stein's method and Brownian motion embeddings. The coverage of martingales, probability bounds, Markov chains, and renewal theory focuses on applications in applied probability, where a number of recently developed results from the literature are given.

The book can be used in a flexible fashion: After starting with Chapter 1, you may take the remaining chapters in almost any order, with a few caveats. We hope you enjoy this book.

About Notation

Here we assume the reader is familiar with the mathematical notation used in an elementary probability course. For example, we write $X \sim U(a, b)$ or $X =_d U(a, b)$ to mean that X is a random variable having a uniform distribution between the numbers a and b. We use common abbreviations like $N(\mu, \sigma^2)$ and Poisson(λ) to mean a normal distribution with mean μ and variance σ^2 and a Poisson distribution with parameter λ, respectively. We also write I_A or $I\{A\}$ to denote a random variable that equals 1 if A is true and equals 0 otherwise, and we use the abbreviation iid for random variables to mean independent and identically distributed random variables.

Acknowledgments

We would like to thank the following people for many valuable comments
and suggestions: Jose Blanchet (Stanford University), Joe Blitzstein
(Harvard University), Stephen Herschkorn, Jim Pitman (University of Cal-
ifornia, Berkeley), Amber Puha (California State University, San Marcos),
Ward Whitt (Columbia University), and the students in Statistics 212 at
Harvard University in the Fall 2005 semester.

1

Measure Theory and Laws of Large Numbers

1.1 Introduction

If you're reading this, you've probably already seen many different types of random variables and have applied the usual theorems and laws of probability to them. We will, however, show you there are some seemingly innocent random variables for which none of the laws of probability apply. Measure theory, as it applies to probability, is a theory that carefully describes the types of random variables the laws of probability apply to. This puts the whole field of probability and statistics on a mathematically rigorous foundation.

You are probably familiar with some proof of the famous strong law of large numbers, which asserts that the long-run average of independent and identically distributed (iid) random variables converges to the expected value. One goal of this chapter is to show you a beautiful and more general alternative proof of this result using the powerful ergodic theorem. In order to do this, we will first take you on a brief tour of measure theory and introduce you to the dominated convergence theorem, one of measure theory's most famous results and the key ingredient we need.

In Section 1.2, we construct an event, called a nonmeasurable event, to which the laws of probability don't apply. In Section 1.3, we introduce the notions of countably and uncountably infinite sets and show you how the elements of some infinite sets cannot be listed in a sequence. In Section 1.4, we define a probability space and the laws of probability that apply to them. In Section 1.5, we introduce the concept of a measurable random variable, and in Section 1.6, we introduce the concepts of convergence and limits. In Section 1.7, we define the expected value in terms of the Lebesgue integral. In Section 1.8, we illustrate and prove the dominated convergence theorem, and Section 1.9, we discuss convergence in probability and distribution. Lastly, in Section 1.10, we prove zero-one laws and the ergodic theorem and use these to obtain the strong law of large numbers.

1.2   A Nonmeasurable Event

Consider a circle that has a radius equal to one. We say that two points on the edge of the circle are in the same family if you can go from one point to the other point by taking steps of length one unit around the edge of the circle. By this we mean each step you take moves you an angle of exactly one radian degree around the circle, and you are allowed to keep looping around the circle in either direction.

Suppose each family elects one of its members to be the head of the family. Here is the question: What is the probability a point X selected uniformly at random along the edge of the circle is the head of its family? It turns out this question has no answer.

The first thing to notice is that each family has an infinite number of family members. Because the circumference of the circle is 2π, you can never get back to your starting point by looping around the circle with steps of length one. If it were possible to start at the top of the circle and get back to the top going a steps clockwise and looping around b times, then you would have $a = b2\pi$ for some integers a, b, and hence $\pi = a/(2b)$. This is impossible because it's well-known that π is an irrational number and can't be written as a ratio of integers.

It may seem to you like the probability should either be zero or one, but we will show you why neither answer could be correct. It doesn't even depend on how the family heads are elected. Define the events $A = \{X$ is the head of its family$\}$, $A_i = \{X$ is i steps clockwise from the head of its family$\}$, and $B_i = \{X$ is i steps counterclockwise from the head of its family$\}$.

Because X was uniformly chosen, we must have $P(A) = P(A_i) = P(B_i)$. But because every family has a head, the sum of these probabilities should equal one, or in other words,

$$1 = P(A) + \sum_{i=1}^{\infty}(P(A_i) + P(B_i)).$$

Thus, if $x = P(A)$ we get $1 = x + \sum_{i=1}^{\infty} 2x$, which has no solution where $0 \leq x \leq 1$. This means it's impossible to compute $P(A)$, and the answer is neither zero nor one, nor any other possible number. The event A is called a non-measurable event, because you can't measure its probability in a consistent way.

What's going on here? It turns out that allowing only one head per family, or any finite number of heads, is what makes this event nonmeasurable. If we allowed more than one head per family and gave everyone a 50% chance, independent of all else, of being a head of the family, then we would have no trouble measuring the probability of this event. Or if we let everyone in the top half of the circle be a family head, and again let families have more than one head, the answer would be easy. Later we will give a careful description of what types of events we can actually compute probabilities for.

Being allowed to choose exactly one family head from each family requires a special mathematical assumption called the axiom of choice. This axiom famously can create all sorts of other logical mayhem, such as allowing you to break a sphere into a finite number of pieces and rearrange them into two spheres of the same size (the Banach–Tarski paradox). For this reason, the axiom is controversial and has been the subject of much study by mathematicians.

1.3   Countable and Uncountable Sets

You may now be asking yourself if the existence of a uniform random variable $X \sim U(0,1)$ also contradicts the laws of probability. We know that for all x, $P(X = x) = 0$, but also $P(0 \le X \le 1) = 1$. Doesn't this give a contradiction because

$$P(0 \le X \le 1) = \sum_{x \in [0,1]} P(X = x) = 0?$$

Actually, this is not a contradiction because a summation over an interval of real numbers does not make any sense. Which values of x would you use for the first few terms in the sum? The first term in the sum could use $x = 0$, but it's difficult to decide which value of x to use next.

In fact, infinite sums are defined in terms of a sequence of finite sums:

$$\sum_{i=1}^{\infty} x_i \equiv \lim_{n \to \infty} \sum_{i=1}^{n} x_i,$$

so to have an infinite sum, it must be possible to arrange the terms in a sequence. If an infinite set of items can be arranged in a sequence it is called *countable*; otherwise it is called *uncountable*.

Obviously the integers are countable using the sequence 0, −1, +1, −2, +2, ... The positive rational numbers are also countable if you express them as a ratio of integers and list them in order by the sum of these integers:

$$\frac{1}{1}, \frac{2}{1}, \frac{1}{2}, \frac{3}{1}, \frac{2}{2}, \frac{1}{3}, \frac{4}{1}, \frac{3}{2}, \frac{2}{3}, \frac{1}{4}, \dots$$

The real numbers between zero and one, however, are not countable. Here we will explain why. Suppose somebody thinks they have a method of arranging them into a sequence $x_1, x_2, ...$, where we express them as $x_j = \sum_{i=1}^{\infty} d_{ij} 10^{-i}$ so that $d_{ij} \in \{0, 1, 2, ..., 9\}$ is the ith digit after the decimal place of the jth number in their sequence. Then you can clearly see that the number

$$y = \sum_{i=1}^{\infty} (1 + I\{d_{ii} = 1\}) 10^{-i},$$

where $I\{A\}$ equals one if A is true and zero otherwise is nowhere to be found in their sequence. This is because y differs from x_i in at least the ith decimal place, so it is different from every number in their sequence. Whenever someone tries to arrange the real numbers into a sequence, this shows that they will always be omitting some of the numbers. This proves that the real numbers in any interval are uncountable and that you can't take a sum over all of them.

So it's true with $X \sim U(0, 1)$ that for any countable set A we have $P(X \in A) = \sum_{x \in A} P(X = x) = 0$, but we can't simply sum up the probabilities like this for an uncountable set. There are, however, some examples of uncountable sets A (the Cantor set, for example) that have $P(X \in A) = 0$.

1.4   Probability Spaces

Let Ω be the set of points in a sample space, and let $\mathcal{F}$ be the collection of subsets of Ω for which we can calculate a probability. These subsets are called events and can be viewed as possible things that could happen. If we let P be the function that gives the probability for any event in $\mathcal{F}$, then the triple $(\Omega, \mathcal{F}, P)$ is called a probability space. The collection $\mathcal{F}$ is usually what is called a sigma field (also called a sigma algebra), which we define next.

Definition 1.1 *The collection of sets $\mathcal{F}$ is a sigma field, or a σ field, if it has the following three properties:*

 1. $\Omega \in \mathcal{F}$

 2. $A \in \mathcal{F} \rightarrow A^c \in \mathcal{F}$

 3. $A_1, A_2, ... \in \mathcal{F} \rightarrow \cup_{i=1}^{\infty} A_i \in \mathcal{F}$.

These properties say you can calculate the probability of the whole sample space (Property 1), the complement of any event (Property 2), and the countable union of any sequence of events (Property 3). They also imply that you can calculate the probability of the countable intersection of any sequence of events because $\cap_{i=1}^{\infty} A_i = (\cup_{i=1}^{\infty} A_i^c)^c$.

To specify a σ field, people typically start with a collection of events $\mathcal{A}$ and write $\sigma(\mathcal{A})$ to represent the smallest σ field containing the collection of events $\mathcal{A}$. Thus $\sigma(\mathcal{A})$ is called the σ field "generated" by $\mathcal{A}$. It is uniquely defined as the intersection of all possible sigma fields that contain $\mathcal{A}$, and in Exercise 3 at the end of this chapter, you will show such an intersection is always a sigma field.

Example 1.2 Let $\Omega = \{a, b, c\}$ be the sample space, and let $\mathcal{A} = \{\{a, b\}, \{c\}\}$. Then $\mathcal{A}$ is not a a σ field because $\{a, b, c\} \notin \mathcal{A}$, but $\sigma(\mathcal{A}) = \{\{a, b, c\}, \{a, b\}, \{c\}, \phi\}$, where $\phi = \Omega^c$ is the empty set.

Definition 1.3 *A probability measure P is a function, defined on the sets in a sigma field, which has the following three properties:*

1. $P(\Omega) = 1$, and

2. $P(A) \geq 0$, and

3. $P(\cup_{i=1}^{\infty} A_i) = \sum_{i=1}^{\infty} P(A_i)$ *if* $\forall i \neq j$ *we have* $A_i \cap A_j = \phi$.

These properties imply that probabilities must be between zero and one and say that the probability of a countable union of mutually exclusive events is the sum of the probabilities.

Example 1.4 *Dice.* If you roll a pair of dice, the 36 points in the sample space are $\Omega = \{(1, 1), (1, 2), ..., (5, 6), (6, 6)\}$. We can let $\mathcal{F}$ be the collection of all possible subsets of Ω, and it's easy to see that it is a sigma field. Then we can define

$$P(A) = \frac{|A|}{36},$$

where $|A|$ is the number of sample space points in A. Thus, if $A = \{(1, 1), (3, 2)\}$, then $P(A) = 2/36$, and it's easy to see that P is a probability measure.

Example 1.5 *The unit interval.* Suppose we want to pick a uniform random number between zero and one. Then the sample space equals $\Omega = [0, 1]$, the set of all real numbers between zero and one. We can let $\mathcal{F}$ be the collection of all possible subsets of Ω, and it's easy to see that it is a sigma field. But it turns out that it's not possible to put a probability measure on this sigma field. Because one of the sets in $\mathcal{F}$ would be similar to the set of heads of the family (from the nonmeasurable event example), this event cannot have a probability assigned to it. So this sigma field is not a good one to use in probability.

Example 1.6 *The unit interval again.* Again with $\Omega = [0,1]$, suppose we use the sigma field $\mathcal{F} = \sigma(\{x\}_{x \in \Omega})$, the smallest sigma field generated by all possible sets containing a single real number. This is a nice enough sigma field, but it would never be possible to find the probability for some interval, such as $[0.2, 0.4]$. You can't take a countable union of single real numbers and expect to get an uncountable interval somehow. So this is not a good sigma field to use.

If we want to put a probability measure on the real numbers between zero and one, what sigma field can we use? The answer is the *Borel sigma field* $\mathcal{B}$, the smallest sigma field generated by all intervals of the form $[x, y)$ of real numbers between zero and one: $\mathcal{B} = \sigma([x, y)_{x < y \in \Omega})$. The sets in this sigma field are called Borel sets. We will see that most reasonable sets you would be interested in are Borel sets, although sets similar to the one in the "heads of the family" example are not Borel sets.

We can then use the special probability measure, which is called a *Lebesgue measure* (named after the French mathematician Henri Lebesgue), defined by $P([x, y)) = y - x$, for $0 \leq x \leq y \leq 1$, to give us a uniform distribution. Defining it for just these intervals is enough to uniquely specify the probability of every set in $\mathcal{B}$. (This fact can be shown to follow from Theorem 1.65, which is discussed later). And actually, you can do almost all of probability starting from just a uniform(0,1) random variable, so this probability measure is pretty much all you need.

Example 1.7 If $\mathcal{B}$ is the Borel sigma field on [0,1], is $\{.5\} \in \mathcal{B}$? Yes, because $\{0.5\} = \cap_{i=1}^{\infty}[0.5, 0.5 + 1/i)$. Also note that $\{1\} = [0, 1)^c \in \mathcal{B}$.

Example 1.8 If $\mathcal{B}$ is the Borel sigma field on [0,1], is the set of rational numbers between zero and one $Q \in \mathcal{B}$? The argument from the previous example shows $\{x\} \in \mathcal{B}$ for all x, so each number by itself is a Borel set, and we then get $Q \in \mathcal{B}$ because Q is countable union of such numbers. Also note that this then means $Q^c \in \mathcal{B}$, so the set of irrational numbers is also a Borel set.

There are some Borel sets that can't directly be written as a countable intersection or union of intervals like the preceding, but you usually don't run into them.

From the definition of probability, we can derive many of the famous formulas you may have seen before such as

$$P(A \cup B) = P(A) + P(B) - P(A \cap B),$$

and extending this by induction,

$$P(\cup_{i=1}^n A_i) = \sum_i P(A_i) - \sum_{i<j} P(A_i \cap A_j)$$
$$+ \sum_{i<j<k} P(A_i \cap A_j \cap A_k) \cdots$$
$$\cdots + (-1)^{n+1} P(A_1 \cap A_2 \cdots \cap A_n),$$

where the last formula is usually called the *inclusion–exclusion formula*. Next we give a couple of examples applying these. In these examples, the sample space is finite, and in such cases unless otherwise specified, we assume the corresponding sigma field is the set of all possible subsets of the sample space.

Example 1.9 *Cards.* A deck of n cards is well shuffled many times. (a) What's the probability the cards all get back to their initial positions? (b) What's the probability at least one card is back in its initial position?

Solution Because there are $n!$ different ordering for the cards and all are approximately equally likely after shuffling, the answer to Part (a) is approximately $1/n!$. For the answer to Part (b), let $A_i = \{$card i is back in its initial position$\}$ and let $A = \cup_{i=1}^\infty A_i$ be the event at least one card is back in its initial position. Because $P(A_{i_1} \cap A_{i_2} \cap ... \cap A_{i_k}) = (n-k)!/n!$, and because the number of terms in the kth sum of the inclusion–exclusion formula is $\binom{n}{k}$, we have

$$P(A) = \sum_{k=1}^n (-1)^{k+1} \binom{n}{k} \frac{(n-k)!}{n!}$$
$$= \sum_{k=1}^n \frac{(-1)^{k+1}}{k!}$$
$$\approx 1 - 1/e$$

for large n. ∎

Example 1.10 *Coins.* If a fair coin is flipped n times, what is the chance of seeing at least k heads in row?

Solution We will show you that the answer is

$$\sum_{m=1}^{(n+1)/(k+1)} (-1)^{m+1} \left[\binom{n-mk}{m} 2^{-m(k+1)} + \binom{n-mk}{m-1} 2^{-m(k+1)+1} \right].$$

When we define the event $A_i = \{$a run of a tail immediately followed by k heads in a row starts at flip $i\}$, and $A_0 = \{$the first k flips are heads$\}$, we can use the inclusion–exclusion formula to get this solution because

$$P(\text{at least } k \text{ heads in row}) = P(\cup_{i=0}^{n-k-1} A_i)$$

and

$$P(A_{i_1} A_{i_2} \cdots A_{i_m}) = \begin{cases} 0 & \text{if flips for any events overlap} \\ 2^{-m(k+1)} & \text{otherwise and } i_1 > 0 \\ 2^{-m(k+1)+1} & \text{otherwise and } i_1 = 0 \end{cases}$$

and the number of sets of indices $i_1 < i_2 < \cdots < i_m$, where the runs that do not overlap equal $\binom{n-mk}{m}$ if $i_1 > 0$ (imagine the k heads in each of the m runs are invisible, so this is the number of ways to arrange m tails in $n - mk$ visible flips) and $\binom{n-mk}{m-1}$ if $i_1 = 0$. ∎

An important property of the probability function is that it is a continuous function on the events of the sample space Ω. To make this precise, let $A_n, n \geq 1$ be a sequence of events, and define the event $\liminf A_n$ as

$$\liminf A_n \equiv \cup_{n=1}^{\infty} \cap_{i=n}^{\infty} A_i.$$

Because $\liminf A_n$ consists of all outcomes of the sample space that are contained in $\cap_{i=n}^{\infty} A_i$ for some n, it follows that $\liminf A_n$ consists of all outcomes that are contained in all but a finite number of the events $A_n, n \geq 1$.

Similarly, the event $\limsup A_n$ is defined by

$$\limsup A_n = \cap_{n=1}^{\infty} \cup_{i=n}^{\infty} A_i.$$

Because $\limsup A_n$ consists of all outcomes of the sample space that are contained in $\cup_{i=n}^{\infty} A_i$ for all n, it follows that $\limsup A_n$ consists of all outcomes that are contained in an infinite number of the events $A_n, n \geq 1$. Sometimes the notation $\{A_n \text{ i.o.}\}$ is used to represent $\limsup A_n$, where i.o. stands for infinitely often and means that an infinite number of the events A_n occur.

Note that by their definitions

$$\liminf A_n \subset \limsup A_n.$$

Definition 1.11 *If* $\limsup A_n = \liminf A_n$, *we say that* $\lim_n A_n$ *exists and define it by*

$$\lim_n A_n = \limsup A_n = \liminf A_n.$$

Example 1.12 (a) Suppose that $A_n, n \geq 1$ is an increasing sequence of events, in that $A_n \subset A_{n+1}, n \geq 1$. Then $\cap_{i=n}^{\infty} A_i = A_n$, showing that

$$\liminf A_n = \cup_{n=1}^{\infty} A_n.$$

Also, $\cup_{i=n}^{\infty} A_i = \cup_{i=1}^{\infty} A_i$, showing that

$$\limsup A_n = \cup_{n=1}^{\infty} A_n.$$

Hence,

$$\lim_n A_n = \cup_{i=1}^{\infty} A_i.$$

(b) If $A_n, n \geq 1$ is a decreasing sequence of events, in that $A_{n+1} \subset A_n, n \geq 1$, then it similarly follows that

$$\lim_n A_n = \cap_{i=1}^{\infty} A_i. \quad \blacksquare$$

The following result is known as the *continuity property of probabilities*.

Proposition 1.13 *If* $\lim_n A_n = A$, *then* $\lim_n P(A_n) = P(A)$.

Proof We prove it first for when A_n is either an increasing or decreasing sequence of events. Suppose $A_n \subset A_{n+1}, n \geq 1$. Then, with A_0 defined to be the empty set,

$$\begin{aligned}
P(\lim A_n) &= P(\cup_{i=1}^{\infty} A_i) \\
&= P(\cup_{i=1}^{\infty} A_i (\cup_{j=1}^{i-1} A_j)^c) \\
&= P(\cup_{i=1}^{\infty} A_i A_{i-1}^c) \\
&= \sum_{i=1}^{\infty} P(A_i A_{i-1}^c) \\
&= \lim_{n \to \infty} \sum_{i=1}^{n} P(A_i A_{i-1}^c) \\
&= \lim_{n \to \infty} P(\cup_{i=1}^{n} A_i A_{i-1}^c) \\
&= \lim_{n \to \infty} P(\cup_{i=1}^{n} A_i) \\
&= \lim_{n \to \infty} P(A_n).
\end{aligned}$$

Now, suppose that $A_{n+1} \subset A_n, n \geq 1$. Because A_n^c is an increasing sequence of events, the preceding implies that

$$P(\cup_{i=1}^{\infty} A_i^c) = \lim_{n \to \infty} P(A_n^c),$$

or equivalently,

$$P((\cap_{i=1}^{\infty} A_i)^c) = 1 - \lim_{n \to \infty} P(A_n)$$

or

$$P(\cap_{i=1}^{\infty} A_i) = \lim_{n \to \infty} P(A_n),$$

which completes the proof whenever A_n is a monotone sequence. Now, consider the general case, and let $B_n = \cup_{i=n}^{\infty} A_i$. Noting that $B_{n+1} \subset B_n$, and applying the preceding yields

$$\begin{aligned} P(\limsup A_n) &= P(\cap_{n=1}^{\infty} B_n) \\ &= \lim_{n \to \infty} P(B_n). \end{aligned} \tag{1.1}$$

Also, with $C_n = \cap_{i=n}^{\infty} A_i$,

$$\begin{aligned} P(\liminf A_n) &= P(\cup_{n=1}^{\infty} C_n) \\ &= \lim_{n \to \infty} P(C_n) \end{aligned} \tag{1.2}$$

because $C_n \subset C_{n+1}$. But

$$C_n = \cap_{i=n}^{\infty} A_i \subset A_n \subset \cup_{i=n}^{\infty} A_i = B_n,$$

showing that

$$P(C_n) \leq P(A_n) \leq P(B_n). \tag{1.3}$$

Thus, if $\liminf A_n = \limsup A_n = \lim A_n$, then we obtain from Equations 1.1 and 1.2 that the upper and lower bounds of Equation 1.3 converge to each other in the limit, and this proves the result. ∎

1.5   Random Variables

Suppose you have a function X that assigns a real number to each point in the sample space Ω and you also have a sigma field $\mathcal{F}$. We say that X is an $\mathcal{F}$-measurable random variable if you can compute its entire cumulative distribution function using probabilities of events in $\mathcal{F}$ or, equivalently, that you would know the value of X if you were told which events in $\mathcal{F}$ actually happen. We define the notation $\{X \leq x\} \equiv \{\omega \in \Omega : X(\omega) \leq x\}$, so X is $\mathcal{F}$ measurable if $\{X \leq x\} \in \mathcal{F}$ for all x. This is often written in shorthand notation as $X \in \mathcal{F}$.

Example 1.14 $\Omega = \{a, b, c\}$, $\mathcal{A} = \{\{a, b, c\}, \{a, b\}, \{c\}, \phi\}$, and we define three random variables X, Y, Z as follows:

ω	X	Y	Z
a	1	1	1
b	1	2	7
c	2	2	4

Which of the random variables X, Y, and Z are $\mathcal{A}$ measurable? Because $\{Y \le 1\} = \{a\} \notin \mathcal{A}$, then Y is not $\mathcal{A}$ measurable. For the same reason, Z is not $\mathcal{A}$ measurable. The variable X is $\mathcal{A}$ measurable because $\{X \le 1\} = \{a, b\} \in \mathcal{A}$, and $\{X \le 2\} = \{a, b, c\} \in \mathcal{A}$. In other words, you can always figure out the value of X using just the events in $\mathcal{A}$, but you can't always figure out the values of Y and Z.

Definition 1.15 *For a random variable X we define*

$$\sigma(X) = \sigma(\{X \le x\}, \forall x)$$

to be the sigma field generated by all events of the type $\{\mathbf{X} \le x\}$, where $\sigma(X)$ is the sigma field generated by X.

Alternatively, we can define $\sigma(X)$ as the intersection of all possible sigma fields $\mathcal{F}$ where X is $\mathcal{F}$ measurable; such an uncountable intersection is a sigma field, as in Exercise 3 at the end of this chapter. Intuitively, $\sigma(X)$ contains just enough events to know the value of X when you know which of the events occur.

Definition 1.16 *For random variables X, Y we say that X is Y measurable if $X \in \sigma(Y)$.*

Example 1.17 In the previous example, is $Y \in \sigma(Z)$? Yes, because $\sigma(Z) = \{\{a, b, c\}, \{a\}, \{a, b\}, \{b\}, \{b, c\}, \{c\}, \{c, a\}, \phi\}$, the set of all possible subsets of Ω. Is $X \in \sigma(Y)$? No, because $\{X \le 1\} = \{a, b\} \notin \sigma(Y) = \{\{a, b, c\}, \{b, c\}, \{a\}, \phi\}$.

  To see why $\sigma(Z)$ is as given, note that $\{Z \le 1\} = \{a\}$, $\{Z \le 4\} = \{a, c\}$, $\{Z \le 7\} = \{a, b, c\}$, $\{a\}^c = \{b, c\}$, $\{a, b\}^c = \{c\}$, $\{a\} \cup \{c\} = \{a, c\}$, $\{a, b, c\}^c = \phi$, and $\{a, c\}^c = \{b\}$.

Example 1.18 Suppose X and Y are random variables taking values between zero and one and are measurable with respect to the Borel sigma field $\mathcal{B}$. Is $Z = X + Y$ also measurable with respect to $\mathcal{B}$? Well, we must show that $\{Z \le z\} \in \mathcal{B}$ for all z. We can write

$$\{X + Y > z\} = \cup_{q \in Q}(\{X > q\} \cap \{Y > z - q\}),$$

where Q is the set of rational numbers. Because $\{X > q\} \in \mathcal{B}$, $\{Y > z - q\} \in \mathcal{B}$, and Q is countable, this means that $\{X + Y \le z\} = \{X + Y > z\}^c \in \mathcal{B}$ and thus Z is measurable with respect to $\mathcal{B}$.

Example 1.19 The function $F(x) = P(X \leq x)$ is called the distribution function of the random variable X. If $x_n \downarrow x$ then the sequence of events $A_n = \{X \leq x_n\}$, $n \geq 1$, is a decreasing sequence with a limit that is

$$\lim_n A_n = \cap_n A_n = \{X \leq x\}.$$

Consequently, the continuity property of probabilities yields

$$F(x) = \lim_{n \to \infty} F(x_n),$$

showing that a distribution function is always right continuous. On the other hand, if $x_n \uparrow x$, then the sequence of events $A_n = \{X \leq x_n\}$, $n \geq 1$, is an increasing sequence, implying that

$$\lim_{n \to \infty} F(x_n) = P(\cup_n A_n) = P(X < x) = F(x) - P(X = x).$$

Two events are independent if knowing that one occurs does not change the chance that the other occurs. This is formalized in the following definition.

Definition 1.20 *Sigma fields $\mathcal{F}_1, \ldots, \mathcal{F}_n$ are independent if whenever $A_i \in \mathcal{F}_i$ for $i = 1, \ldots, n$, we have $P(\cap_{i=1}^n A_i) = \prod_{i=1}^n P(A_i)$.*

Using this we say that random variables $X_1, \ldots, X_n$ are *independent* if the sigma fields $\sigma(X_1), \ldots, \sigma(X_n)$ are independent, and we say events $A_1, \ldots, A_n$ are independent if $I_{A_1}, \ldots, I_{A_n}$ are independent random variables.

Remark 1.21 One interesting property of independence is that it's possible that events A, B, C are not independent even if each pair of the events are independent. For example, if we make three independent flips of a fair coin and let A represent the event exactly one head comes up in the first two flips, let B represent the event exactly one head comes up in the last two flips, and let C represent the event exactly one head comes up among the first and last flip. Then each event has probability $1/2$, the intersection of each pair of events has probability $1/4$, but we have $P(ABC) = 0$.

In our next example, we derive a formula for the distribution of the convolution of geometric random variables.

Example 1.22 Suppose we have n coins that we toss in sequence, moving from one coin to the next in line each time a head appears. That is, we continue using a coin until it lands heads, and then we switch to the next one. Let X_i denote the number of flips made with coin i. Assuming that all coin flips are independent and that each lands heads with probability p, we know from our first course in probability that X_i is a geometric random variable with parameter p and that the total number of flips made has a negative binomial distribution with probability mass function

$$P(X_1 + \cdots + X_n = m) = \binom{m-1}{n-1} p^n (1-p)^{m-n}, \quad m \geq n.$$

The probability mass function of the total number of flips when each coin has a different probability of landing heads is easily obtained using the following proposition.

Proposition 1.23 *If $X_1, \ldots, X_n$ are independent geometric random variables with parameters $p_1, \ldots, p_n$, where $p_i \neq p_j$ if $i \neq j$, then, with $q_i = 1 - p_i$, for $k \geq n - 1$*

$$P(X_1 + \cdots + X_n > k) = \sum_{i=1}^{n} q_i^k \prod_{j \neq i} \frac{p_j}{p_j - p_i}.$$

Proof We will prove $A_{k,n} = P(X_1 + \cdots + X_n > k)$ is as given using induction on $k + n$. Because $A_{1,1} = q_1$, we will assume as our induction hypothesis that $A_{i,j}$ is as given previously for all $i + j < k + n$. Then, depending on whether or not the event $\{X_n > 1\}$ occurs, we get

$$A_{k,n} = q_n A_{k-1,n} + p_n A_{k-1,n-1}$$

$$= q_n \sum_{i=1}^{n} q_i^{k-1} \prod_{j \neq i} \frac{p_j}{p_j - p_i} + p_n \sum_{i=1}^{n-1} q_i^{k-1} \frac{p_n - p_i}{p_n} \prod_{j \neq i} \frac{p_j}{p_j - p_i}$$

$$= \sum_{i=1}^{n} q_i^k \prod_{j \neq i} \frac{p_j}{p_j - p_i},$$

which completes the proof by induction. ∎

1.6 Convergence, Limits, sup, and inf

A sequence of real numbers $x_1, x_2, \ldots$ *converges to a limit* x, and we write this as $\lim_{n \to \infty} x_n = x$ or $\lim_n x_n = x$ or $x_n \to x$ if for any $\epsilon > 0$ the values in the sequence beyond some point are all within ϵ of x. We write $x_n \uparrow x$ if $x_n \to x$ and the sequence is nondecreasing, and we write $x_n \downarrow x$ if $x_n \to x$ and the sequence is nonincreasing.

If X_n is a sequence of random variables and we write $X_n \to X$, we mean that if we observe the sequence and then consider it as a sequence of real numbers, we will always have $X_n \to X$.

Example 1.24 If $x_n = n/(n+1)$ for $n = 1, 2, \ldots$ then we have $x_n \uparrow 1$. This is because x_n is nondecreasing, and given any $\epsilon > 0$, we can let $n = 1/\epsilon$ and $1 - x_i = 1/(i+1) < \epsilon$ when when $i > n$.

Example 1.25 If $x_n = n/(n+1)$ when n is even and $x_n = 0$ when n is odd, we say that the sequence has no limit. Because for $n \geq 1$ we have $x_{2n} \geq 2/3$ and $x_{2n+1} = 0$, when $\epsilon = 1/3$ we can never find an n such that all the values beyond the nth value are less than ϵ from the same number.

If x_i for $i \in S$ are real numbers with indices in a set S we write

$$x = \sup_{i \in S} x_i$$

if $x_i \leq x$ for all i and for any $y < x$ there is some $i \in S$ such that $x_i > y$. We say that x is the *supremum* of the set $\{x_i : i \in S\}$, which means it is the smallest possible upper bound for the set. Here S may be either a countable or an uncountable set. We also define the infimum of a set as the largest possible lower bound so that if

$$x = \inf_{i \in S} x_i$$

it means $x_i \geq x$ for all i and for any $y > x$ there is some $i \in S$ such that $x_i < y$.

Example 1.26 If $S = \{1, 2, \ldots\}$ and $x_i = i$, we have that $\sup_{i \in S} x_i = \infty$ and $\inf_{i \in S} x_i = 1$. Also note that there is no maximum value of x_i, so $\max_{i \in S} x_i$ does not exist.

Every set of real numbers has a supremum and an infimum, although these may not actually be in the set. Infinite sets may not have a maximum or minimum value within them, although finite sets always do.

1.7   Expected Value

A random variable X is *continuous* if there is a function f, called its density function, so $P(X \leq x) = \int_{-\infty}^{x} f(t)dt$ for all x. A random variable is *discrete* if it can only take a countable number of different values. In elementary textbooks, you usually see two separate definitions for expected value:

$$E[X] = \begin{cases} \sum_i x_i P(X = x_i) & \text{if } X \text{ is discrete} \\ \int x f(x)dx & \text{if } X \text{ is continuous with density } f. \end{cases}$$

But it's possible to have a random variable that is neither continuous nor discrete. For example, with $U \sim U(0,1)$, the variable $X = UI_{U>0.5}$ is neither continuous nor discrete. It's also possible to have a sequence of continuous random variables that converges to a discrete random variable or vice versa. For example, if $X_n = U/n$, then each X_n is a continuous random variable, but $\lim_{n\to\infty} X_n$ is a discrete random variable (which equals zero). This means it would be better to have a single more general definition that covers all types of random variables. We introduce this next.

A *simple* random variable is one that can take on only a finite number of different possible values, and its expected value is defined as in the first paragraph in this section for discrete random variables. Using these, we next define the expected value of a more general nonnegative random variable. We will later define it for general random variables X by expressing it as the difference of two nonnegative random variables $X = X^+ - X^-$, where $x^+ = \max(0, x)$ and $x^- = \max(-x, 0)$.

Definition 1.27 *If $X \geq 0$, then we define*

$$E[X] \equiv \sup_{\text{all simple variables } Y \leq X} E[Y].$$

We write $Y \leq X$ for random variables X, Y to mean $P(Y \leq X) = 1$; this is sometimes written as "$Y \leq X$ almost surely" and abbreviated "$Y \leq X$ a.s." For example, if X is nonnegative and $a \geq 0$, then $Y = aI_{X \geq a}$ is a simple random variable such that $Y \leq X$. And by taking a supremum over all simple variables, we of course mean the simple random variables must be measurable with respect to some given sigma field. Given a nonnegative random variable X, one concrete choice of simple variables is the sequence $Y_n = \min(\lfloor 2^n X \rfloor / 2^n, n)$, where $\lfloor x \rfloor$ denotes the integer portion of x. In Exercise 18 at the end of this chapter, we ask you to show that $Y_n \uparrow X$ and $E[X] = \lim_n E[Y_n]$.

Another consequence of the definition of expected value is that if $Y \leq X$, then $E[Y] \leq E[X]$.

Example 1.28 *Markov's inequality.* Suppose $X \geq 0$. Then, for any $a > 0$ we have that $a I_{X \geq a} \leq X$. Therefore, $E[a I_{X \geq a}] \leq E[X]$ or, equivalently,

$$P(X \geq a) \leq E[X]/a,$$

which is known as Markov's inequality.

Example 1.29 *Chebyshev's inequality.* A consequence of Markov's inequality is that for $a > 0$

$$P(|X| \geq a) = P(X^2 \geq a^2) \leq E[X^2]/a^2,$$

a result known as Chebyshev's inequality.

Given any random variable $X \geq 0$ with $E[X] < \infty$, and any $\epsilon > 0$, we can find a simple random variable Y with $E[X] - \epsilon \leq E[Y] \leq E[X]$. Our definition of the expected value also gives what is called the Lebesgue integral of X with respect to the probability measure P and is sometimes denoted $E[X] = \int X dP$.

So far we have only defined the expected value of a nonnegative random variable. For the general case, we first define $X^+ = XI_{X \geq 0}$ and $X^- = -XI_{X < 0}$ so that we can define $E[X] = E[X^+] - E[X^-]$, with the convention that $E[X]$ is undefined if $E[X^+] = E[X^-] = \infty$.

Remark 1.30 The definition of expected value covers random variables that are neither continuous nor discrete, but if X is continuous with density function f, it is equivalent to the familiar definition $E[X] = \int x f(x) dx$. For example, when $0 \leq X \leq 1$ the definition of the Riemann integral in terms of Riemann sums implies, with $\lfloor x \rfloor$ denoting the integer portion of x,

$$
\int_0^1 x f(x) dx = \lim_{n \to \infty} \sum_{i=0}^{n-1} \int_{i/n}^{(i+1)/n} x f(x) dx
$$

$$
\leq \lim_{n \to \infty} \sum_{i=0}^{n-1} \frac{i+1}{n} P\left(i/n \leq X \leq \frac{i+1}{n}\right)
$$

$$
= \lim_{n \to \infty} \sum_{i=0}^{n-1} i/n P\left(i/n \leq X \leq \frac{i+1}{n}\right)
$$

$$
= \lim_{n \to \infty} E[\lfloor nX \rfloor / n]
$$

$$
\leq E[X],
$$

where the last line follows because $\lfloor nX \rfloor / n \leq X$ is a simple random variable.

Using that the density function g of $1 - X$ is $g(x) = f(1 - x)$, we obtain

$$
1 - E[X] = E[1 - X]
$$

$$
\geq \int_0^1 x f(1 - x) dx
$$

$$
= \int_0^1 (1 - x) f(x) dx
$$

$$
= 1 - \int_0^1 x f(x) dx.
$$

Remark 1.31 At this point, you may think it might be possible to express any random variable as sums or mixtures of discrete and continuous random variables, but this is not true. Let $X \sim U(0, 1)$ be a uniform random variable, and let $d_i \in \{0, 1, 2, ..., 9\}$ be the ith digit in its decimal expansion so that $X = \sum_{i=1}^{\infty} d_i 10^{-i}$. The random variable $Y = \sum_{i=1}^{\infty} \min(1, d_i) 10^{-i}$ is not discrete and has no intervals over which it is continuous. This variable Y can take any value (between zero and one) having a decimal expansion that uses only the digits 0 and 1, which are a set of values C called a *Cantor set*. Because C contains no intervals, Y is not continuous. And Y is not discrete because C is uncountable; every real number between zero and one, using its base two expansion, corresponds to a distinct infinite sequence of binary digits.

Another interesting fact about a Cantor set is, although C is uncountable, $P(X \in C) = 0$. Let C_i be the set of real numbers between zero and one that have a decimal expansion using only the digits 0 and 1 up to the ith decimal place. Then it's easy to see that $P(X \in C_i) = 0.2^i$ and because $P(X \in C) \leq P(X \in C_i) = 0.2^i$ for any i, we must have $P(X \in C) = 0$. The set C is called an uncountable set having measure zero.

Proposition 1.32 *If $E|X|, E|Y| < \infty$ then (a) $E[aX + b] = aE[X] + b$ for constants a, b, and (b) $E[X + Y] = E[X] + E[Y]$.*

Proof In this proof we assume $X, Y \geq 0$, $a > 0$, and $b = 0$. The general cases will follow using $E[X + Y] = E[X^+ + Y^+] - E[X^- + Y^-]$,

$$E[b + X] = \sup_{Y \leq b+X} E[Y] = \sup_{Y \leq X} E[b + Y] = b + \sup_{Y \leq X} E[Y] = b + E[X],$$

and $-aX + b = a(-X) + b$.

For Part (a) if X is simple we have

$$E[aX] = \sum_x axP(X = x) = aE[X],$$

and because for every simple variable $Z \leq X$ there corresponds another simple variable $aZ \leq aX$, and vice versa, we get

$$E[aX] = \sup_{aZ \leq aX} E[aZ] = \sup_{Z \leq X} aE[Z] = aE[X],$$

where the supremums are over simple random variables.

For Part (b) if X, Y are simple we have

$$
\begin{aligned}
E[X + Y] &= \sum_z z P(X + Y = z) \\
&= \sum_z z \sum_{x,y: x+y=z} P(X = x, Y = y) \\
&= \sum_z \sum_{x,y: x+y=z} (x + y) P(X = x, Y = y) \\
&= \sum_{x,y} (x + y) P(X = x, Y = y) \\
&= \sum_{x,y} x P(X = x, Y = y) + \sum_{x,y} y P(X = x, Y = y) \\
&= \sum_x x P(X = x) + \sum_y y P(Y = y) \\
&= E[X] + E[Y],
\end{aligned}
$$

and applying this in the following second line, we get

$$
\begin{aligned}
E[X] + E[Y] &= \sup_{A \leq X, B \leq Y} E[A] + E[B] \\
&= \sup_{A \leq X, B \leq Y} E[A + B] \\
&\leq \sup_{A \leq X + Y} E[A] \\
&= E[X + Y],
\end{aligned}
$$

where the supremums are over simple random variables. We then use this inequality in the following third line:

$$
\begin{aligned}
E[\min(X + Y, n)] &= 2n - E[2n - \min(X + Y, n)] \\
&\leq 2n - E[n - \min(X, n) + n - \min(Y, n)] \\
&\leq 2n - E[n - \min(X, n)] - E[n - \min(Y, n)] \\
&= E[\min(X, n)] + E[\min(Y, n)] \\
&\leq E[X] + E[Y],
\end{aligned}
$$

and we use Part (a) in the first and fourth lines and $\min(X + Y, n) \leq \min(X, n) + \min(Y, n)$ in the second line.

This means for any given simple $Z \leq X + Y$ we can pick n larger than the maximum value of Z so that $E[Z] \leq E[\min(X + Y, n)] \leq E[X] + E[Y]$, and taking the supremum over all simple $Z \leq X + Y$ gives $E[X + Y] \leq E[X] + E[Y]$ and the result is proved. ∎

Proposition 1.33 *If X is a nonnegative integer valued random variable, then*

$$E[X] = \sum_{n=0}^{\infty} P(X > n).$$

Proof Because $E[X] = p_1 + 2p_2 + 3p_3 + 4p_4 \ldots$ (see Exercise 7 at the end of this chapter), where $p_i = P(X = i)$, we rewrite this as

$$
\begin{aligned}
E[X] = p_1 + &\ p_2 + p_3 + p_4 \ \cdots \\
+ &\ p_2 + p_3 + p_4 \ \cdots \\
&\ \quad\ p_3 + p_4 \ \cdots \\
&\ \qquad\quad\ p_4 \ \cdots\,.
\end{aligned}
$$

Notice that the columns equal $p_1, 2p_2, 3p_3, \ldots$, respectively, whereas the rows equal $P(X > 0), P(X > 1), P(X > 2), \ldots$, respectively. ∎

Example 1.34 With $X_1, X_2 \ldots$ independent $U(0,1)$ random variables, compute the expected value of

$$N = \min\left\{ n : \sum_{i=1}^{n} X_i > 1 \right\}.$$

Solution Using $E[N] = \sum_{n=0}^{\infty} P(N > n)$, and noting that

$$P(N > 0) = P(N > 1) = 1,$$

and

$$
\begin{aligned}
P(N > n) &= \int_0^1 \int_0^{1-x_1} \int_0^{1-x_1-x_2} \cdots \int_0^{1-x_1-x_2-\cdots-x_{n-1}} dx_n \cdots dx_1 \\
&= 1/n!,
\end{aligned}
$$

we get $E[N] = e$. ∎

1.8 Almost Sure Convergence and the Dominated Convergence Theorem

For a sequence of nonrandom real numbers, recall that we write $x_n \to x$ or $\lim_{n \to \infty} x_n = x$ if for any $\varepsilon > 0$ there exists a value n such that $|x_m - x| < \varepsilon$ for all $m > n$. Intuitively, this means eventually the sequence never leaves an arbitrarily small neighborhood around x. It doesn't simply mean that you can always find terms in the sequence that are arbitrarily close to x, but rather it means that eventually *all* terms in the sequence become arbitrarily close to x. When $x_n \to \infty$, it means that for any $k > 0$ there exists a value n such that $x_m > k$ for all $m > n$.

The sequence of random variables $X_n, n \geq 1$, is said to converge *almost surely* to the random variable X, written as $X_n \longrightarrow_{as} X$, or $\lim_{n \to \infty} X_n = X$ a.s., if with

$$\lim_n X_n = X.$$

The following proposition presents an alternative characterization of almost sure convergence.

Proposition 1.35 $X_n \longrightarrow_{as} X$ *if and only if for any $\epsilon > 0$*

$$P(|X_n - X| < \epsilon \text{ for all } n \geq m) \to 1 \quad as \quad m \to \infty.$$

Proof Suppose first that $X_n \longrightarrow_{as} X$. Fix $\epsilon > 0$, and for $m \geq 1$, define the event

$$A_m = \{|X_n - X| < \epsilon \text{ for all } n \geq m\}.$$

Because $A_m, m \geq 1$, is an increasing sequence of events, the continuity property of probabilities yields that

$$
\begin{aligned}
\lim_m P(A_m) &= P(\lim_m A_m) \\
&= P(|X_n - X| < \epsilon \text{ for all } n \text{ sufficiently large}) \\
&\geq P(\lim_n X_n = X) \\
&= 1.
\end{aligned}
$$

To go the other way, assume that for any $\epsilon > 0$

$$P(|X_n - X| < \epsilon \text{ for all } n \geq m) \to 1 \quad as \quad m \to \infty.$$

Let $\epsilon_i, i \geq 1$, be a decreasing sequence of positive numbers that converge to 0, and let

$$A_{m.i} = \{|X_n - X| < \epsilon_i \text{ for all } n \geq m\}.$$

Because $A_{m.i} \subset A_{m+1.i}$ and, by assumption, $\lim_m P(A_{m,i}) = 1$, it follows from the continuity property that

$$1 = P(\lim_{m \to \infty} A_{m.i}) = P(B_i),$$

where $B_i = \{|X_n - X| < \epsilon_i \text{ for all } n \text{ sufficiently large}\}$. But $B_i, i \geq 1$, is a decreasing sequence of events, so invoking the continuity property once again yields

$$1 = \lim_{i \to \infty} P(B_i) = P(\lim_i B_i),$$

which proves the result because

$$\lim_i B_i = \{\text{for all } i, |X_n - X| < \epsilon_i \text{ for all } n \text{ sufficiently large}\}$$
$$= \{\lim_n X_n = X\}.$$

∎

Remark 1.36 The reason for the word almost in "almost surely" is because $P(A) = 1$ doesn't necessarily mean that A^c is the empty set. For example, if $X \sim U(0,1)$, we know that $P(X \neq 1/3) = 1$ even though $\{X = 1/3\}$ is a possible outcome.

The dominated convergence theorem is one of the fundamental building blocks of all limit theorems in probability. It tells you something about what happens to the expected value of random variables in a sequence if the random variables are converging almost surely. Many limit theorems in probability involve an almost surely converging sequence, and being able to accurately say something about the expected value of the limiting random variable is important.

Given a sequence of random variables $X_1, X_2, \ldots$, it may seem to you at first thought that $X_n \to X$ a.s. should imply $\lim_{n \to \infty} E[X_n] = E[X]$. This is sometimes called *interchanging limit and expectation*, because $E[X] = E[\lim_{n \to \infty} X_n]$. But this interchange is not always valid, and the next example illustrates this.

Example 1.37 Suppose $U \sim U(0,1)$ and $X_n = nI_{n<1/U}$. Regardless of what U turns out to be, as soon as n gets larger than $1/U$, we see that the terms X_n in the sequence will all equal zero. This means $X_n \to 0$ a.s., but at the same time we have $E[X_n] = nP(U < 1/n) = n/n = 1$ for all n, and thus $\lim_{n \to \infty} E[X_n] = 1$. Interchanging limit and expectation is not valid in this case.

What's going wrong here? In this case, X_n can increase beyond any level as n gets larger and larger, and this can cause problems with the expected value. The dominated convergence theorem says that if X_n is always bounded in absolute value by some other random variable with finite mean, then we can interchange limit and expectation. We will first state the theorem, give some examples, and then give a proof. The proof is a nice illustration of the definition of expected value.

Proposition 1.38 *The dominated convergence theorem. Suppose $X_n \to X$ a.s., and there is a random variable Y with $E[Y] < \infty$ such that $|X_n| < Y$ for all n. Then*

$$E[\lim_{n \to \infty} X_n] = \lim_{n \to \infty} E[X_n].$$

This is often used in the form where Y is a nonrandom constant, and then it's called the *bounded convergence theorem*. Before we prove it, we first give a couple of examples and illustrations.

Example 1.39 Suppose $U \sim U(0,1)$ and $X_n = U/n$. It's easy to see that $X_n \to 0$ a.s., and the theorem would tell us that $E[X_n] \to 0$. In fact, in this case we can easily calculate $E[X_n] = \frac{1}{2n} \to 0$. The theorem applies using $Y = 1$ because $|X_n| < 1$.

Example 1.40 With $X \sim N(0,1)$, let $X_n = \min(X, n)$, and notice $X_n \to X$ almost surely. Because $X_n < |X|$, we can apply the theorem using $Y = |X|$ to tell us $E[X_n] \to E[X]$.

Example 1.41 Suppose $X \sim N(0,1)$ and let $X_n = X I_{X \geq -n} - n I_{X < -n}$. Again $X_n \to X$, so using $Y = |X|$ the theorem tells us $E[X_n] \to E[X]$.

Proof *Proof of the dominated convergence theorem.* To be able to directly apply the definition of expected value, in this proof we assume $X_n \geq 0$. To prove the general result, we can apply the same argument to $X_n + Y \geq 0$ with the bound $|X_n + Y| < 2Y$.

Our approach will be to show that for any $\varepsilon > 0$ we have, for all sufficiently large n, both (a) $E[X_n] \geq E[X] - 3\varepsilon$ and (b) $E[X_n] \leq E[X] + 3\varepsilon$. Because ε is arbitrary, this will prove the theorem.

First, let $N_\varepsilon = \min\{n : |X_i - X| < \varepsilon$ for all $i \geq n\}$, and note that $X_n \longrightarrow_{as} X$ implies that $P(N_\varepsilon < \infty) = 1$. To Part (a), note first that for any m

$$X_n + \varepsilon \geq \min(X, m) - m I_{N_\varepsilon > n}.$$

The preceding is true when $N_\varepsilon > n$ because in this case the right-hand side is nonpositive; it is also true when $N_\varepsilon \leq n$ because in this case $X_n + \varepsilon \geq X$. Thus,

$$E[X_n] + \varepsilon \geq E[\min(X, m)] - m P(N_\varepsilon > n).$$

Now, $|X| \leq Y$ implies that $E[X] \leq E[Y] < \infty$. Consequently, using the definition of $E[X]$, we can find a simple random variable $Z \leq X$ with $E[Z] \geq E[X] - \varepsilon$. Because Z is simple, we can then pick m large enough so $Z \leq \min(X, m)$, and thus

$$E[\min(X, m)] \geq E[Z] \geq E[X] - \varepsilon.$$

Then $N_\varepsilon < \infty$ implies, by the continuity property, that $m P(N_\varepsilon > n) < \varepsilon$ for sufficiently large n. Combining this with the preceding shows that for sufficiently large n

$$E[X_n] + \varepsilon \geq E[X] - 2\varepsilon,$$

which is Part (a).

For Part (b), apply Part (a) to the sequence of nonnegative random variables $Y - X_n$, which converges almost surely to $Y - X$ with a bound $|Y - X_n| < 2Y$. We get $E[Y - X_n] \geq E[Y - X] - 3\varepsilon$, and rearranging and subtracting $E[Y]$ from both sides gives Part (b). $\blacksquare$

Remark 1.42 Part (a) in the proof holds for nonnegative random variables even without the upper bound Y and under the weaker assumption that $\inf_{m>n} X_m \to X$ as $n \to \infty$. This result is usually referred to as *Fatou's lemma*, which states that for any $\epsilon > 0$ we have $E[X_n] \geq E[X] - \epsilon$ for sufficiently large n, or equivalently that $\inf_{m>n} E[X_m] \geq E[X] - \epsilon$ for sufficiently large n. This result is usually denoted as $\liminf_{n\to\infty} E[X_n] \geq E[\liminf_{n\to\infty} X_n]$.

A result called the *monotone convergence theorem* can also be proved.

Proposition 1.43 *The monotone convergence theorem. If*

$$0 \leq X_n \uparrow X,$$

then $E[X_n] \uparrow E[X]$.

Proof If $E[X] < \infty$, we can apply the dominated convergence theorem using the bound $|X_n| < X$.

Consider now the case where $E[X] = \infty$. For any m, we have $\min(X_n, m) \to \min(X, m)$. Because $E[\min(X, m)] < \infty$, it follows by the dominated convergence theorem that

$$\lim_n E[\min(X_n, m)] = E[\min(X, m)].$$

But because $E[X_n] \geq E[\min(X_n, m)]$, this implies

$$\lim_n E[X_n] \geq \lim_{m\to\infty} E[\min(X, m)].$$

Because $E[X] = \infty$, it follows that for any K there is a simple random variable $A \leq X$ such that $E[A] > K$. Because A is simple, $A \leq \min(X, m)$ for sufficiently large m. Thus, for any K

$$\lim_{m\to\infty} E[\min(X, m)] \geq E[A] > K,$$

proving that $\lim_{m\to\infty} E[\min(X, m)] = \infty$ and completing the proof. $\blacksquare$

We now present a couple of corollaries of the monotone convergence theorem.

Corollary 1.44 *If* $X_i \geq 0$, *then* $E[\sum_{i=1}^{\infty} X_i] = \sum_{i=1}^{\infty} E[X_i]$.

Proof

$$\sum_{i=1}^{\infty} E[X_i] = \lim_n \sum_{i=1}^{n} E[X_i]$$

$$= \lim_n E\left[\sum_{i=1}^{n} X_i\right]$$

$$= E\left[\sum_{i=1}^{\infty} X_i\right],$$

where the final equality follows from the monotone convergence theorem because $\sum_{i=1}^{n} X_i \uparrow \sum_{i=1}^{\infty} X_i$. ∎

Corollary 1.45 *If X and Y are independent, then*

$$E[XY] = E[X]E[Y].$$

Proof Suppose first that X and Y are simple. Then we can write

$$X = \sum_{i=1}^{n} x_i I_{\{X=x_i\}}, \quad Y = \sum_{j=1}^{m} y_j I_{\{Y=y_j\}}.$$

Thus,

$$E[XY] = E\left[\sum_i \sum_j x_i y_j I_{\{X=x_i, Y=y_j\}}\right]$$

$$= \sum_i \sum_j x_i y_j E[I_{\{X=x_i, Y=y_j\}}]$$

$$= \sum_i \sum_j x_i y_j P(X = x_i, Y = y_j)$$

$$= \sum_i \sum_j x_i y_j P(X = x_i) P(Y = y_j)$$

$$= E[X]E[Y].$$

Next, suppose X, Y are general nonnegative random variables. For any n, define the simple random variables

$$X_n = \begin{cases} k/2^n, & \text{if } \frac{k}{2^n} < X \leq \frac{k+1}{2^n}, \ k = 0, \ldots, n2^n - 1. \\ n, & \text{if } X > n \end{cases}$$

Define random variables Y_n in a similar fashion, and note that

$$X_n \uparrow X, \ Y_n \uparrow Y, \ X_n Y_n \uparrow XY.$$

Hence, by the monotone convergence theorem,

$$E[X_n Y_n] \to E[XY].$$

But X_n and Y_n are simple, so

$$E[X_n Y_n] = E[X_n]E[Y_n] \to E[X]E[Y],$$

with the convergence again following by the monotone convergence theorem. Thus, $E[XY] = E[X]E[Y]$ when X and Y are nonnegative. The general case follows by writing $X = X^+ - X^-$, $Y = Y^+ - Y^-$, using

$$E[XY] = E[X^+Y^+] - E[X^+Y^-] - E[X^-Y^+] + E[X^-Y^-]$$

and applying the result to each of the four preceding expectations. ∎

1.9 Convergence in Probability and in Distribution

In this section, we introduce two forms of convergence that are weaker than almost sure convergence. However, before giving their definitions, we will start with a useful result, known as the *Borel–Cantelli lemma*.

Proposition 1.46 *If $\sum_j P(A_j) < \infty$, then $P(\limsup A_k) = 0$.*

Proof Suppose $\sum_j P(A_j) < \infty$. Now,

$$P(\limsup A_k) = P(\cap_{n=1}^{\infty} \cup_{i=n}^{\infty} A_i).$$

Hence, for any n

$$P(\limsup A_k) \leq P(\cup_{i=n}^{\infty} A_i)$$
$$\leq \sum_{i=n}^{\infty} P(A_i),$$

and the result follows by letting $n \to \infty$. ∎

Remark Because $\sum_n I_{A_n}$ is the number of events $A_n, n \geq 1$, that occur, the Borel–Cantelli theorem states that if the expected number of events $A_n, n \geq 1$, that occur is finite, then the probability that an infinite number of them occur is zero. Thus, the Borel–Cantelli lemma is equivalent to the rather intuitive result that if there is a positive probability that an infinite number of the events A_n occur, and then the expected number of them that occur is infinite.

The converse of the Borel–Cantelli lemma requires that the indicator variables for each pair of events be negatively correlated.

Proposition 1.47 *Let the events $A_i, i \geq 1$, be such that*

$$\text{Cov}(I_{A_i}, I_{A_j}) = E[I_{A_i} I_{A_j}] - E[I_{A_i}] E[I_{A_j}] \leq 0, \ i \neq j.$$

If $\sum_{i=1}^{\infty} P(A_i) = \infty$, then $P(\limsup A_i) = 1$.

Proof Let $N_n = \sum_{i=1}^{n} I_{A_i}$ be the number of the events $A_1, \ldots, A_n$ that occur, and let $N = \sum_{i=1}^{\infty} I_{A_i}$ be the total number of events that occur. Let $m_n = E[N_n] = \sum_{i=1}^{n} P(A_i)$, and note that $\lim_n m_n = \infty$. Using the formula for the variance of a sum of random variables learned in your first course in probability, we have

$$
\begin{aligned}
\text{Var}(N_n) &= \sum_{i=1}^{n} \text{Var}(I_{A_i}) + 2 \sum_{i<j} \text{Cov}(I_{A_i}, I_{A_j}) \\
&\leq \sum_{i=1}^{n} \text{Var}(I_{A_i}) \\
&= \sum_{i=1}^{n} P(A_i)[1 - P(A_i)] \\
&\leq m_n.
\end{aligned}
$$

Now, by Chebyshev's inequality, for any $x < m_n$

$$
\begin{aligned}
P(N_n < x) &= P(m_n - N_n > m_n - x) \\
&\leq P(|N_n - m_n| > m_n - x) \\
&\leq \frac{\text{Var}(N_n)}{(m_n - x)^2} \\
&\leq \frac{m_n}{(m_n - x)^2}.
\end{aligned}
$$

Hence, for any x, $\lim_{n \to \infty} P(N_n < x) = 0$. Because $P(N < x) \leq P(N_n < x)$, this implies that

$$P(N < x) = 0.$$

Consequently, by the continuity property of probabilities,

$$
\begin{aligned}
0 &= \lim_{k \to \infty} P(N < k) \\
&= P\left(\lim_k \{N < k\}\right) \\
&= P(\cup_k \{N < k\}) \\
&= P(N < \infty).
\end{aligned}
$$

Hence, with a probability of one, an infinite number of the events A_i occur. ∎

Example 1.48 Consider independent flips of a coin that lands heads with probability $p > 0$. For fixed k, let B_n be the event that flips $n, n+1, \ldots, n+k-1$ all land heads. Because the events $B_n, n \geq 1$, are positively correlated, we cannot directly apply the converse to the Borel–Cantelli lemma to obtain that, with a probability of 1; an infinite number of them occur. However, by letting A_n be the event that flips $nk+1, \ldots, nk+k$ all land heads, then because the set of flips these events refer to are nonoverlapping, it follows that they are independent. Because $\sum_n P(A_n) = \sum_n p^k = \infty$, we obtain from Borel–Cantelli that $P(\limsup A_n) = 1$. But $\limsup A_n \subset \limsup B_n$, so the preceding yields the result $P(\limsup B_n) = 1$. ∎

Remark 1.49 The converse of the Borel–Cantelli lemma is usually stated as requiring the events $A_i, i \geq 1$, to be independent. Our weakening of this condition can be useful, as the next example shows.

Example 1.50 Consider an infinite collection of balls that are numbered $0, 1, \ldots$ and an infinite collection of boxes also numbered $0, 1, \ldots$. Suppose that ball $i, i \geq 0$, is to be put in box $i + X_i$, where $X_i, i \geq 0$, are iid with probability mass function

$$P(X_i = j) = p_j \qquad \sum_{j \geq 0} p_j = 1.$$

Suppose also that the X_i are not deterministic, so $p_j < 1$ for all $j \geq 0$. If A_j denotes the event that box j remains empty, then

$$
\begin{aligned}
P(A_j) &= P(X_j \neq 0, X_{j-1} \neq 1, \ldots, X_0 \neq j) \\
&= P(X_0 \neq 0, X_1 \neq 1, \ldots, X_j \neq j) \\
&\geq P(X_i \neq i, \text{for all } i \geq 0).
\end{aligned}
$$

But

$$
\begin{aligned}
P(X_i \neq i, &\text{ for all } i \geq 0) \\
&= 1 - P(\cup_{i \geq 0}\{X_i = i\}) \\
&= 1 - p_0 - \sum_{i \geq 1} P(X_0 \neq 0, \ldots, X_{i-1} \neq i-1, X_i = i) \\
&= 1 - p_0 - \sum_{i \geq 1} p_i \prod_{j=0}^{i-1}(1 - p_j).
\end{aligned}
$$

Now, there is at least one pair $k < i$ such that $p_i p_k \equiv p > 0$. Hence, for that pair

$$p_i \prod_{j=0}^{i-1}(1 - p_j) \leq p_i(1 - p_k) = p_i - p,$$

implying that

$$P(A_j) \geq P(X_i \neq i, \text{for all } i \geq 0) \geq p > 0.$$

Hence, $\sum_j P(A_j) = \infty$. Conditional on box j being empty, each ball becomes more likely to be put in box $i, i \neq j$, so for $i < j$,

$$P(A_i|A_j) = \prod_{k=0}^{i} P(X_k \neq i - k|A_j)$$

$$= \prod_{k=0}^{i} P(X_k \neq i - k|X_k \neq j - k)$$

$$\leq \prod_{k=0}^{i} P(X_k \neq i - k)$$

$$= P(A_i),$$

which is equivalent to $\text{Cov}(I_{A_i}, I_{A_j}) \leq 0$. Hence, by the converse of the Borel–Cantelli lemma we can conclude that, with a probability of one, there will be an infinite number of empty boxes.

We say that the sequence of random variables $X_n, n \geq 1$, *converges in probability* to the random variable X, written $X_n \longrightarrow_p X$, if for any $\epsilon > 0$

$$P(|X_n - X| > \epsilon) \to 0 \quad \text{as} \quad n \to \infty.$$

An immediate corollary of Proposition 1.35 is that almost sure convergence implies convergence in probability. The following example shows that the converse is not true.

Example 1.51 Let $X_n, n \geq 1$ be independent random variables such that

$$P(X_n = 1) = 1/n = 1 - P(X_n = 0).$$

For any $\epsilon > 0$, $P(|X_n| > \epsilon) = 1/n \to 0$; hence, $X_n \longrightarrow_p 0$. However, because $\sum_{n=1}^{\infty} P(X_n = 1) = \infty$, it follows from the converse to the Borel–Cantelli lemma that $X_n = 1$ for infinitely many values of n, showing that the sequence does not converge almost surely to zero.

Let F_n be the distribution function of X_n, and let F be the distribution function of X. We say that X_n converges in distribution to X if

$$\lim_{n \to \infty} F_n(x) = F(x)$$

for all x at which F is continuous. (That is, convergence is required at all x for which $P(X = x) = 0$.)

To understand why convergence in distribution only requires that $F_n(x) \to F(x)$ at points of continuity of F, rather than at all values x, let X_n be uniformly distributed on $(0, 1/n)$. Then, it seems reasonable to suppose that X_n converges in distribution to the random variable X that is identically zero. However,

$$F_n(x) = \begin{cases} 0, & \text{if } x < 0 \\ nx, & \text{if } 0 \le x \le 1/n, \\ 1, & \text{if } x > 1/n \end{cases}$$

whereas the distribution function of X is

$$F(x) = \begin{cases} 0, & \text{if } x < 0 \\ 1, & \text{if } x \ge 0. \end{cases}$$

Thus, $\lim_n F_n(0) = 0 \ne F(0) = 1$. On the other hand, for all points of continuity of F (that is, for all $x \ne 0$), we have that $\lim_n F_n(x) = F(x)$, so with the definition given, it is indeed true that $X_n \longrightarrow_d X$.

We now show that convergence in probability implies convergence in distribution.

Proposition 1.52

$$X_n \longrightarrow_p X \quad \Rightarrow \quad X_n \longrightarrow_d X.$$

Proof Suppose that $X_n \longrightarrow_p X$. Let F_n be the distribution function of $X_n, n \ge 1$, and let F be the distribution function of X. Now, for any $\epsilon > 0$

$$\begin{aligned} F_n(x) &= P(X_n \le x, \, X \le x + \epsilon) + P(X_n \le x, \, X > x + \epsilon) \\ &\le F(x + \epsilon) + P(|X_n - X| > \epsilon), \end{aligned}$$

where the preceding used

$$X_n \le x, \, X > x + \epsilon \Rightarrow |X_n - X| > \epsilon.$$

Letting n go to infinity yields, upon using $X_n \longrightarrow_p X$,

$$\limsup_{n \to \infty} F_n(x) \le F(x + \epsilon). \tag{1.4}$$

Similarly,

$$\begin{aligned} F(x - \epsilon) &= P(X \le x - \epsilon, \, X_n \le x) + P(X \le x - \epsilon, \, X_n > x) \\ &\le F_n(x) + P(|X_n - X| > \epsilon). \end{aligned}$$

Letting $n \to \infty$ gives

$$F(x - \epsilon) \leq \liminf_{n \to \infty} F_n(x). \tag{1.5}$$

Combining Equations 1.4 and 1.5 shows that

$$F(x - \epsilon) \leq \liminf_{n \to \infty} F_n(x) \leq \limsup_{n \to \infty} F_n(x) \leq F(x + \epsilon).$$

Letting $\epsilon \to 0$ shows that if x is a continuity point of F then

$$F(x) \leq \liminf_{n \to \infty} F_n(x) \leq \limsup_{n \to \infty} F_n(x) \leq F(x),$$

and the result is proved.   ∎

Proposition 1.53   *If* $X_n \longrightarrow_d X$, *then*

$$E[g(X_n)] \to E[g(X)]$$

for any bounded continuous function g.

To focus on the essentials, we will present a proof of Proposition 1.53 when all the random variables X_n and X are continuous. Before doing so, we will prove a couple of lemmas.

Lemma 1.54   *Let G be the distribution function of a continuous random variable, and let $G^{-1}(x) \equiv \inf \{t : G(t) \geq x\}$, be its inverse function. If U is a uniform $(0, 1)$ random variable, then $G^{-1}(U)$ has distribution function G.*

Proof Because

$$\inf \{t : G(t) \geq U\} \leq x \Leftrightarrow G(x) \geq U$$

implies

$$P(G^{-1}(U) \leq x) = P(G(x) \geq U) = G(x),$$

we get the result. ∎

Lemma 1.55   *Let $X_n \longrightarrow_d X$, where X_n is continuous with distribution function F_n, $n \geq 1$, and X is continuous with distribution function F. If $F_n(x_n) \to F(x)$, where $0 < F(x) < 1$ then $x_n \to x$.*

Proof Suppose there is an $\epsilon > 0$ such that $x_n \leq x - \epsilon$ for infinitely many n. If so, then $F_n(x_n) \leq F_n(x - \epsilon)$ for infinitely many n, implying that

$$F(x) = \liminf_n F_n(x_n) \leq \lim_n F_n(x - \epsilon) = F(x - \epsilon),$$

which is a contradiction. We arrive at a similar contradiction upon assuming there is an $\epsilon > 0$ such that $x_n \geq x + \epsilon$ for infinitely many n. Consequently, we can conclude that for any $\epsilon > 0$, $|x_n - x| > \epsilon$ for only a finite number of n, thus proving the lemma. ∎

Proof of Proposition 1.53 Let U be a uniform $(0,1)$ random variable, and set $Y_n = F_n^{-1}(U)$, $n \geq 1$, and $Y = F^{-1}(U)$. Note that from Lemma 1.54 it follows that Y_n has distribution F_n and Y has distribution F. Because

$$F_n(F_n^{-1}(u)) = u = F(F^{-1}(u)),$$

it follows from Lemma 1.55 that $F_n^{-1}(u) \to F^{-1}(u)$ for all u. Thus, $Y_n \longrightarrow_{as} Y$. By continuity, this implies that $g(Y_n) \longrightarrow_{as} g(Y)$, and because g is bounded, the dominated convergence theorem yields that $E[g(Y_n)] \to E[g(Y)]$. But X_n and Y_n both have distribution F_n, whereas X and Y both have distribution F, so $E[g(Y_n)] = E[g(X_n)]$ and $E[g(Y)] = E[g(X)]$. ∎

Remark 1.56 The key to our proof of Proposition 1.53 was showing that, if $X_n \longrightarrow_d X$, we can define random variables $Y_n, n \geq 1$, and Y such that Y_n has the same distribution as X_n for each n, and Y has the same distribution as X, and are such that $Y_n \longrightarrow_{as} Y$. This result (which is true without the continuity assumptions we made) is known as *Skorokhod's representation theorem*.

Skorokhod's representation and the dominated convergence theorem immediately yield the following.

Corollary 1.57 *If $X_n \longrightarrow_d X$ and there exists a constant $M < \infty$ such that $|X_n| < M$ for all n, then*

$$\lim_{n \to \infty} E[X_n] = E[X].$$

Proof Let F_n denote the distribution of X_n, $n \geq 1$, and F that of X. Let U be a uniform $(0,1)$ random variable, and for $n \geq 1$, set $Y_n = F_n^{-1}(U)$, and $Y = F^{-1}(U)$. Note that the hypotheses of the corollary imply that $Y_n \longrightarrow_{as} Y$ and, because $F_n(M) = 1 = 1 - F_n(-M)$, also that $|Y_n| \leq M$. Thus, by the dominated convergence theorem

$$E[Y_n] \to E[Y],$$

which proves the result because Y_n has distribution F_n, and Y has distribution F. ∎

Proposition 1.53 can also be used to give a simple proof of Weierstrass' approximation theorem.

Corollary 1.58 *Weierstrass' approximation theorem. Any continuous function f defined on the interval $[0, 1]$ can be expressed as a limit of polynomial functions. Specifically, if*

$$B_n(t) = \sum_{i=0}^{n} f(i/n) \binom{n}{i} t^i (1 - t)^{n-i},$$

then $f(t) = \lim_{n \to \infty} B_n(t)$.

Proof Let $X_i, i \geq 1$, be a sequence of iid random variables such that

$$P(X_i = 1) = t = 1 - P(X_i = 0).$$

Because $E[\frac{X_1 + \cdots + X_n}{n}] = t$, it follows from Chebyshev's inequality that for any $\epsilon > 0$

$$P\left(\left| \frac{X_1 + \cdots + X_n}{n} - t \right| > \epsilon \right) \leq \frac{\mathrm{Var}([X_1 + \cdots + X_n]/n)}{\epsilon^2} = \frac{p(1 - p)}{n\epsilon^2}.$$

Thus, $\frac{X_1 + \cdots + X_n}{n} \to_p t$, implying that $\frac{X_1 + \cdots + X_n}{n} \to_d t$. Because f is a continuous function on a closed interval, it is bounded and so Proposition 1.53 yields

$$E\left[f\left(\frac{X_1 + \cdots + X_n}{n} \right) \right] \to f(t).$$

But $X_1 + \cdots + X_n$ is a binomial (n, t) random variable; thus,

$$E\left[f\left(\frac{X_1 + \cdots + X_n}{n} \right) \right] = B_n(t),$$

and the proof is complete. ∎

1.10 Law of Large Numbers and Ergodic Theorem

Definition 1.59 *For a sequence of random variables $X_1, X_2, \ldots$ the* tail sigma field $\mathcal{T}$ *is defined as*

$$\mathcal{T} = \bigcap_{n=1}^{\infty} \sigma(X_n, X_{n+1}, \ldots).$$

Events $A \in \mathcal{T}$ are called tail events.

Although it may seem as though there are no events remaining in the preceding intersection, there are a lot of examples of interesting tail events. Intuitively, with a tail event you can ignore any finite number of the variables and still be able to tell whether or not the event occurs. Next are some examples.

Example 1.60 Consider a sequence of random variables $X_1, X_2, \ldots$ having tail sigma field $\mathcal{T}$ and satisfying $|X_i| < \infty$ for all i. For the event $A_x = \{\lim_{n\to\infty} \frac{1}{n} \sum_{i=1}^{n} X_i = x\}$, it's easy to see that $A_x \in \mathcal{T}$ because to determine if A_x happens you can ignore any finite number of the random variables; their contributions end up becoming negligible in the limit.

For the event $B_x = \{\sup_i X_i = x\}$, it's easy to see that $B_\infty \in \mathcal{T}$ because it depends on the long-run behavior of the sequence. Note that $B_7 \notin \mathcal{T}$ because it depends, for example, on whether or not $X_1 \leq 7$.

Example 1.61 Consider a sequence of random variables $X_1, X_2, \ldots$ having tail sigma field $\mathcal{T}$, but this time let it be possible for $X_i = \infty$ for some i. For the event $A_x = \{\lim_{n\to\infty} \frac{1}{n} \sum_{i=1}^{n} X_i = x\}$, we now have $A_x \notin \mathcal{T}$ because any variable along the way that equals infinity will affect the limit.

Remark 1.62 The previous two examples also motivate the subtle difference between $X_i < \infty$ and $X_i < \infty$ almost surely. The former means it's impossible to see $X_5 = \infty$, and the latter only says it has probability zero. An event that has probability zero could still be a possible occurrence. For example, if X is a uniform random variable between zero and one, we can write $X \neq 0.2$ almost surely even though it is possible to see $X = 0.2$.

One approach for proving an event always happens is to first prove that its probability must either be zero or one, and then rule out zero as a possibility. This first type of result is called a *zero-one law*, because we are proving the chance must either be zero or one. A nice way to do this is to show an event A is independent of itself, and hence $P(A) = P(A \cap A) = P(A)P(A)$, and thus $P(A) = 0$ or 1. We use this approach next to prove a famous zero-one law for independent random variables, and we will use this in our proof of the law of large numbers.

First, we need the following definition. Events with probability either zero or one are called *trivial* events, and a sigma field is called trivial if every event in it is trivial.

Theorem 1.63 *Kolmogorov's Zero-One Law. A sequence of independent random variables has a trivial tail sigma field.*

Before we give a proof we need the following result. To show that a random variable Y is independent of an infinite sequence of random variables

$X_1, X_2, ...$, it suffices to show that Y is independent of $X_1, X_2, ..., X_n$ for every finite $n < \infty$. In elementary courses, this result is often given as a definition, but it can be justified using measure theory in the next proposition. We define $\sigma(X_i, i \in A) \equiv \sigma(\cup_{i \in A} \sigma(X_i))$ to be the smallest sigma field generated by the collection of random variables $X_i, i \in A$.

Proposition 1.64 *Consider the random variables Y and $X_1, X_2, ...$, where $\sigma(Y)$ is independent of $\sigma(X_1, X_2, ..., X_n)$ for every $n < \infty$. Then $\sigma(Y)$ is independent of $\sigma(X_1, X_2, ...)$.*

Before we prove this proposition, we show how this implies Kolmogorov's zero-one law

Proof *Proof of Kolmogorov's zero-one law.* We will argue that any event $A \in \mathcal{T}$ is independent of itself, and thus $P(A) = P(A \cap A) = P(A)P(A)$ and so $P(A) = 0$ or 1. Note that the tail sigma field $\mathcal{T}$ is independent of $\sigma(X_1, X_2, ..., X_n)$ for every $n < \infty$ (because $\mathcal{T} \subseteq \sigma(X_{n+1}, X_{n+2}, ...)$), so by the previous proposition, it is also independent of $\sigma(X_1, X_2, ...)$. Thus, because $\mathcal{T} \subseteq \sigma(X_1, X_2, ...)$, it also is independent of $\mathcal{T}$. ∎

Now we prove the proposition.

Proof *Proof of Proposition 1.64.* Pick any $A \in \sigma(Y)$. You might at first think that $\mathcal{H} \equiv \cup_{n=1}^{\infty} \sigma(X_1, X_2, ..., X_n)$ is the same as $\mathcal{F} \equiv \sigma(X_1, X_2, ...)$, and then the theorem would follow immediately because by assumption A is independent of any event in $\mathcal{H}$. But it is not true that $\mathcal{H}$ and $\mathcal{F}$ are the same; $\mathcal{H}$ may not even be a sigma field. Also, the tail sigma field $\mathcal{T}$ is a subset of $\mathcal{F}$ but not necessarily of $\mathcal{H}$. It is, however, true that $\mathcal{F} \subseteq \sigma(\mathcal{H})$ (in fact, it turns out that $\sigma(\mathcal{H}) = \mathcal{F}$) because $\sigma(X_1, X_2, ...) \equiv \sigma(\cup_{n=1}^{\infty} \sigma(X_n))$ and $\cup_{n=1}^{\infty} \sigma(X_n) \subseteq \mathcal{H}$. We will use $\mathcal{F} \subseteq \sigma(\mathcal{H})$ later.

Define the collection of events $\mathcal{G}$ to contain any $B \in \mathcal{F}$, where for every $\epsilon > 0$ we can find a corresponding approximating event $C \in \mathcal{H}$ where $P(B \cap C^c) + P(B^c \cap C) \leq \epsilon$. Because A is independent of any event $C \in \mathcal{H}$, we can see that A must also be independent of any event $B \in \mathcal{G}$ because, using the corresponding approximating event C for any desired $\epsilon > 0$,

$$
\begin{aligned}
P(A \cap B) &= P(A \cap B \cap C) + P(A \cap B \cap C^c) \\
&\leq P(A \cap C) + P(B \cap C^c) \\
&\leq P(A)P(C) + \epsilon \\
&= P(A)(P(C \cap B) + P(C \cap B^c)) + \epsilon \\
&\leq P(A)P(B) + 2\epsilon
\end{aligned}
$$

and

$$1 - P(A \cap B) = P(A^c \cup B^c)$$
$$= P(A^c) + P(A \cap B^c)$$
$$= P(A^c) + P(A \cap B^c \cap C) + P(A \cap B^c \cap C^c)$$
$$\leq P(A^c) + P(B^c \cap C) + P(A \cap C^c)$$
$$\leq P(A^c) + \epsilon + P(A)P(C^c)$$
$$= P(A^c) + \epsilon + P(A)(P(C^c \cap B) + P(C^c \cap B^c))$$
$$\leq P(A^c) + 2\epsilon + P(A)P(B^c)$$
$$= 1 + 2\epsilon - P(A)P(B),$$

which when combined gives

$$P(A)P(B) - 2\epsilon \leq P(A \cap B) \leq P(A)P(B) + 2\epsilon.$$

Because ϵ is arbitrary, this shows $\sigma(Y)$ is independent of $\mathcal{G}$. We obtain the proposition by showing $\mathcal{F} \subseteq \sigma(\mathcal{H}) \subseteq \mathcal{G}$ and thus that $\sigma(Y)$ is independent of $\mathcal{F}$, as follows. First note we immediately have $\mathcal{H} \subseteq \mathcal{G}$, and thus $\sigma(\mathcal{H}) \subseteq \sigma(\mathcal{G})$, and we will be finished if we can show $\sigma(\mathcal{G}) = \mathcal{G}$.

To show that $\mathcal{G}$ is a sigma field, clearly $\Omega \in \mathcal{G}$ and $B^c \in \mathcal{G}$ whenever $B \in \mathcal{G}$. Next let $B_1, B_2, ...$ be events in $\mathcal{G}$. To show that $\cup_{i=1}^{\infty} B_i \in \mathcal{G}$, pick any $\epsilon > 0$ and let C_i be the corresponding approximating events that satisfy $P(B_i \cap C_i^c) + P(B_i^c \cap C_i) < \epsilon/2^{i+1}$. Then pick n so that

$$\sum_{i>n} P(B_i \cap B_{i-1}^c \cap B_{i-2}^c \cap \cdots \cap B_1^c) < \epsilon/2.$$

In the following, we use the approximating event $C \equiv \cup_{i=1}^n C_i \in \mathcal{H}$ to get

$$P(\cup_i B_i \cap C^c) + P((\cup_i B_i)^c \cap C)$$

$$\leq P\left(\bigcup_{i=1}^n B_i \cap C^c\right) + \epsilon/2 + P\left(\left(\bigcup_{i=1}^n B_i\right)^c \cap C\right)$$

$$\leq \sum_i P(B_i \cap C_i^c) + P(B_i^c \cap C_i) + \epsilon/2$$

$$\leq \sum_i \epsilon/2^{i+1} + \epsilon/2$$

$$= \epsilon,$$

and thus $\cup_{i=1}^{\infty} B_i \in \mathcal{G}$. ∎

A more powerful theorem, called the extension theorem, can be used to prove Kolmogorov's zero-one law. We state it without proof.

Theorem 1.65 *The extension theorem. Suppose you have random variables $X_1, X_2, ...$, and you consistently define probabilities for all events in $\sigma(X_1, X_2, ..., X_n)$ for every n. This implies a unique value of the probability of any event in $\sigma(X_1, X_2, ...)$.*

Remark 1.66 To see how this implies Kolmogorov's zero-one law, specify probabilities under the assumption that A is independent of any event $B \in \cup_{n=1}^{\infty} \mathcal{F}_n$. The extension theorem will say that A is independent of $\sigma(\cup_{n=1}^{\infty} \mathcal{F}_n)$.

We will prove the law of large numbers using the more powerful ergodic theorem. This means we will show that the long-run average for a sequence of random variables converges to the expected value under more general conditions then just for independent random variables. We will define these more general conditions next.

Given a sequence of random variables $X_1, X_2, \ldots$, suppose (for simplicity and without loss of generality) that there is a one-to-one correspondence between events of the form $\{X_1 = x_1, X_2 = x_2, X_3 = x_3 \ldots\}$ and elements of the sample space Ω. An event A is called an *invariant event* if the occurrence of

$$\{X_1 = x_1, X_2 = x_2, X_3 = x_3 \ldots\} \in A$$

implies both

$$\{X_1 = x_2, X_2 = x_3, X_3 = x_4 \ldots\} \in A$$

and

$$\{X_1 = x_0, X_2 = x_1, X_3 = x_2 \ldots\} \in A.$$

In other words, an invariant event is not affected by shifting the sequence of random variables to the left or right. For example, $A = \{\sup_{n \geq 1} X_n = \infty\}$ is an invariant event if $X_n < \infty$ for all n because $\sup_{n \geq 1} X_n = \infty$ implies both $\sup_{n \geq 1} X_{n+1} = \infty$ and $\sup_{n \geq 1} X_{n-1} = \infty$.

On the other hand, the event $A = \{\lim_n X_{2n} = 0\}$ is not invariant because if a sequence $x_2, x_4, \ldots$ converges to zero it doesn't necessarily mean that $x_1, x_3, \ldots$ converges to zero. Consider the example where $P(X_1 = 1) = 1/2 = 1 - P(X_1 = 0)$ and $X_n = 1 - X_{n-1}$ for $n > 1$. In this case, either $X_{2n} = 0$ and $X_{2n-1} = 1$ for all $n \geq 1$ or $X_{2n} = 1$ and $X_{2n-1} = 0$ for all $n \geq 1$, so $\{\lim_n X_{2n} = 0\}$ and $A = \{\lim_n X_{2n-1} = 0\}$ cannot occur together.

It can be shown (see Exercise 22 at the end of this chapter) that the set of invariant events makes up a sigma field, called the *invariant sigma field*, and is a subset of the tail sigma field. A sequence of random variables $X_1, X_2, \ldots$ is called *ergodic* if it has a trivial invariant sigma field and is called *stationary* if the random variables $(X_1, X_2, \ldots, X_n)$ have the same joint distribution as the random variables $(X_k, X_{k+1}, \ldots, X_{n+k-1})$ for every n, k.

We are now ready to state the ergodic theorem, and an immediate corollary will be the strong law of large numbers.

Theorem 1.67 *The ergodic theorem. If the sequence $X_1, X_2, \ldots$ is stationary and ergodic with $E|X_1| < \infty$, then $\frac{1}{n} \sum_{i=1}^{n} X_i \to E[X_1]$ almost surely.*

Because a sequence of iid random variables is clearly stationary and, by Kolmogorov's zero-one law, ergodic, we get the strong law of large numbers as an immediate corollary.

Corollary 1.68 *The strong law of large numbers. If $X_1, X_2, \ldots$ are iid with $E|X_1| < \infty$, then $\frac{1}{n} \sum_{i=1}^{n} X_i \to E[X_1]$ almost surely.*

Proof *Proof of the ergodic theorem.* Given $\varepsilon > 0$, let $Y_i = X_i - E[X_1] - \varepsilon$ and $M_n = \max(0, Y_1, Y_1 + Y_2, \ldots, Y_1 + Y_2 + \cdots + Y_n)$. Because $\frac{1}{n} \sum_{i=1}^{n} Y_i \le \frac{1}{n} M_n$, we will first show that $M_n/n \to 0$ almost surely, and then the theorem will follow after repeating the whole argument applied instead to $Y_i = -X_i + E[X_1] - \varepsilon$.

Letting $M'_n = \max(0, Y_2, Y_2 + Y_3, \ldots, Y_2 + Y_3 + \cdots + Y_{n+1})$ and using stationarity in the last equality, we have

$$
\begin{aligned}
E[M_{n+1}] &= E[\max(0, Y_1 + M'_n)] \\
&= E[M'_n + \max(-M'_n, Y_1)] \\
&= E[M_n] + E[\max(-M'_n, Y_1)],
\end{aligned}
$$

and because $M_n \le M_{n+1}$ implies $E[M_n] \le E[M_{n+1}]$, we can conclude $E[\max(-M'_n, Y_1)] \ge 0$ for all n.

Because $\{M_n/n \to 0\}$ is an invariant event, by ergodicity it must have probability either zero or one. If we were to assume the probability is zero, then $M_{n+1} \ge M_n$ would imply $M_n \to \infty$ and also $M'_n \to \infty$, and thus $\max(-M'_n, Y_1) \to Y_1$. The dominated convergence theorem using the bound $|\max(-M'_n, Y_1)| \le |Y_1|$ would then give $E[\max(-M'_n, Y_1)] \to E[Y_1] = -\varepsilon$, which would then contradict the previous conclusion that $E[\max(-M'_n, Y_1)] \ge 0$ for all n. This contradiction means we must have $M_n/n \to 0$ almost surely, and the theorem is proved. ∎

1.11 Exercises

1. For $n = 1, 2, \ldots$, let $x_n = (-n)^{-n}$. What can you say about $\sup_n x_n$, $\inf_n x_n$, $\max_n x_n$, $\min_n x_n$, and $\lim_n x_n$?

2. Given a sigma field $\mathcal{F}$, if $A_i \in \mathcal{F}$ for all $1 \le i \le n$, is $\cap_{i=1}^{n} A_i \in \mathcal{F}$?

3. Suppose $\mathcal{F}_i, i = 1, 2, 3, \ldots$ are sigma fields. (a) Is $\cap_{i=1}^{\infty} \mathcal{F}_i$ necessarily always a sigma field? Explain. (b) Does your reasoning in (a) also apply to the intersection of an uncountable number of sigma fields? (c) Is $\cup_{i=1}^{\infty} \mathcal{F}_i$ necessarily always a σ field? Explain.

4. (a) Suppose $\Omega = \{1, 2, \ldots, n\}$. How many different sets will there be in the sigma field generated by starting with the individual elements in Ω? (b) Is it possible for a sigma field to have a countably infinite number of different sets in it? Explain.

5. Show that if X and Y are real-valued random variables measurable with respect to some given sigma field, then so is XY with respect to the same sigma field.

6. If X is a random variable, is it possible for the cumulative distribution function (CDF) $F(x) = P(X \leq x)$ to be discontinuous at a countably infinite number of values of x? Is it possible for it to be discontinuous at an uncountably infinite number of values of x? Explain.

7. Show that $E[X] = \sum_i x_i P(X = x_i)$ if X can only take a countably infinite number of different possible values.

8. Prove that if $X \geq 0$ and $E[X] < \infty$, then $\lim_{n \to \infty} E[X I_{X > n}] = 0$.

9. Assume $X \geq 0$ is a random variable, but don't necessarily assume that $E[1/X] < \infty$. Show that $\lim_{n \to \infty} E[\frac{n}{X} I_{X > n}] = 0$ and $\lim_{n \to \infty} E[\frac{1}{nX} I_{X > n}] = 0$.

10. Use the definition of expected value in terms of simple variables to prove that if $X \geq 0$ and $E[X] = 0$ then $X = 0$ almost surely.

11. Show that if $X_n \longrightarrow_d c$ then $X_n \longrightarrow_p c$.

12. Show that if $E[g(X_n)] \to E[g(X)]$ for all bounded, continuous functions g then $X_n \longrightarrow_d X$.

13. If $X_1, X_2, \ldots$ are nonnegative random variables with the same distribution (but the variables are not necessarily independent) and $E[X_1] < \infty$, prove that $\lim_{n \to \infty} E[\max_{i < n} X_i / n] = 0$.

14. For random variables $X_1, X_2, \ldots$, let $\mathcal{T}$ be the tail sigma field, and let $S_n = \sum_{i=1}^{n} X_i$. (a) Is $\{\lim_{n \to \infty} S_n / n > 0\} \in \mathcal{T}$? (b) Is $\{\lim_{n \to \infty} S_n > 0\} \in \mathcal{T}$?

15. If $X_1, X_2, \ldots$ are nonnegative iid random variables with $P(X_i > 0) > 0$, show that $P(\sum_{i=1}^{\infty} X_i = \infty) = 1$.

16. Suppose $X_1, X_2, \ldots$ are continuous iid random variables and
$$Y_n = I_{\{X_n > \max_{i < n} X_i\}}.$$
(a) Argue that Y_i is independent of Y_j for $i \neq j$. (b) What is $P(\sum_{i=1}^{\infty} Y_i < \infty)$? (c) What is $P(\sum_{i=1}^{\infty} Y_i Y_{i+1} < \infty)$?

17. Suppose there is a single server and the ith customer to arrive requires the server spend U_i time serving them, the time between their arrival and the next customer's arrival is V_i, and $X_i = U_i - V_i$ are iid with mean μ. (a) If Q_{n+1} is the amount of time the $(n + 1)$ customer must wait before being served, explain why $Q_{n+1} = \max(Q_n + X_n, 0)$ $= \max(0, X_n, X_n + X_{n-1}, \ldots, X_n + \cdots + X_1)$. (b) Show $P(Q_n \to \infty) = 1$ if $\mu > 0$.

18. Given a nonnegative random variable X, define the sequence of random variables $Y_n = \min(\lfloor 2^n X \rfloor / 2^n, n)$, where $\lfloor x \rfloor$ denotes the integer portion of x. Show that $Y_n \uparrow X$ and $E[X] = \lim_n E[Y_n]$.

19. Show that for any monotone functions f and g if X, Y are independent random variables then so are $f(X), g(Y)$.

20. Let $X_1, X_2, \ldots$ be random variables with $X_i < \infty$ and suppose $\sum_n P(X_n > 1) < \infty$. Compute $P(\sup_n X_n < \infty)$.

21. Suppose $X_n \to_p X$ and that there is a random variable Y with $E[Y] < \infty$ such that $|X_n| < Y$ for all n. Show $E[\lim_{n \to \infty} X_n] = \lim_{n \to \infty} E[X_n]$.

22. For random variables $X_1, X_2, \ldots$, let $\mathcal{T}$ and $\mathcal{I}$ be the set of tail events and the set of invariant events, respectively. Show that $\mathcal{I}$ and $\mathcal{T}$ are both sigma fields.

23. A ring is hanging from the ceiling by a string. Someone will cut the ring in two positions chosen uniformly at random on the circumference, and this will break the ring into two pieces. Player I gets the piece that falls to the floor, and player II gets the piece that stays attached to the string. Whoever gets the bigger piece wins. Does either player have an advantage here? Explain.

24. A box contains four marbles. One marble is red, and each of the other three marbles is either yellow or green, but you have no idea exactly how many of each color there are or if the other three marbles are all the same color or not. (a) Someone chooses one marble at random from the box, and if you can correctly guess the color, you will win $1,000. What color would you guess? Explain. (b) If this game is to be played four times using the same box of marbles (and the marble drawn each time is placed back in the box), what guesses would you make if you had to make all four guesses ahead of time? Explain.

25. For a sequence of iid continuous random variables $X_1, X_2, \ldots$, let $N = \inf\{n \geq 2 : X_{n+1} > X_n\}$ be the first time the next variable is larger than its immediate predecessor. Compute $E[N]$.

26. Is it possible to pick a random positive integer uniformly at random? Is it possible to pick a positive real number uniformly at random? Explain why or why not.

27. In a group of n people, what is the expected number of distinct birthdays?

28. If a fair coin is flipped n times, what is the expected number of runs of k heads in a row if overlapping runs are each counted separately? What is the expected number of times a run of at least k heads appears in n flips, without counting overlapping runs?

2

Stein's Method and Central Limit Theorems

2.1 Introduction

You are probably familiar with the central limit theorem, which says that the sum of a large number of independent random variables follows roughly a normal distribution. Most proofs presented for this celebrated result generally involve properties of the characteristic function $\phi(t) = E[e^{itX}]$ for a random variable X, the proofs of which are nonprobabilistic and often somewhat mysterious to the uninitiated.

One goal of this chapter is to present a beautiful alternative proof of a version of the central limit theorem using a powerful technique called Stein's method. This technique also amazingly can be applied in settings with dependent variables and gives an explicit bound on the error of the normal approximation; such a bound is difficult to derive using other methods. The technique also can be applied to other distributions, the Poisson and geometric distributions included. We first embark on a brief tour of Stein's method applied in the relatively simpler settings of the Poisson and geometric distributions, and then we move to the normal distribution. As a first step, we introduce the concept of a coupling, one of the key ingredients we need.

In Section 2.2, we introduce the concept of coupling and show how it can be used to bound the error when approximating one distribution with another distribution, and in Section 2.3, we prove a theorem by Le Cam that gives a bound on the error of the Poisson approximation for independent events. In Section 2.4, we introduce the Stein–Chen method, which can give bounds on the error of the Poisson approximation for events with dependencies, and in Section 2.5, we illustrate how the method can be adapted to the setting of the geometric distribution. In Section 2.6, we demonstrate Stein's method applied to the normal distribution, obtain a bound on the error of the normal approximation for the sum of independent variables, and use this to prove a version of the central limit theorem. Lastly, in Section 2.7 we demonstrate Stein's method applied to exponential distribution and use it to approximate the sum of geometric number of independent random variables.

2.2   Coupling

One of the most interesting properties of expected value is that $E[X - Y] = E[X] - E[Y]$ even if the variables X and Y are highly dependent on each other. A useful strategy for estimating $E[X] - E[Y]$ is to create a dependency between X and Y, which simplifies estimating $E[X - Y]$. Such a dependency between two random variables is called a *coupling*.

Definition 2.1 *The pair* $(\widehat{X}, \widehat{Y})$ *is a coupling of the random variables* (X, Y) *if* $\widehat{X} =_d X$ *and* $\widehat{Y} =_d Y$.

Example 2.2 Suppose X, Y and U are $U(0,1)$ random variables. Then both (U, U) and $(U, 1 - U)$ are couplings of (X, Y).

A random variable X is said to be stochastically smaller than Y, also written as $X \leq_{st} Y$, if

$$P(X \leq x) \geq P(Y \leq x), \forall x.$$

Note that if $X \leq Y$ almost surely then $X \leq_{st} Y$. Under this condition, we can create a coupling where one variable is always less than the other.

Proposition 2.3 *If* $X \leq_{st} Y$, *it is almost surely possible to construct a coupling* $(\widehat{X}, \widehat{Y})$ *of* (X, Y) *with* $\widehat{X} \leq \widehat{Y}$.

Proof With $F(t) = P(X \leq t)$, $G(t) = P(Y \leq t)$ and $F^{-1}(x) \equiv \inf\{t : F(t) \geq x\}$ and $G^{-1}(x) \equiv \inf\{t : G(t) \geq x\}$, let $U \sim U(0,1)$, $\widehat{X} = F^{-1}(U)$, and $\widehat{Y} = G^{-1}(U)$. Because $F(t) \geq G(t)$ implies $F^{-1}(x) \leq G^{-1}(x)$, we have $\widehat{X} \leq \widehat{Y}$. And because

$$\inf\{t : F(t) \geq U\} \leq x \Leftrightarrow F(x) \geq U$$

implies

$$P(F^{-1}(U) \leq x) = P(F(x) \geq U) = F(x),$$

we get $\widehat{X} =_d X$ and $\widehat{Y} =_d Y$ after applying the same argument to G. ∎

Example 2.4 If $X \sim N(0,1)$ and $Y \sim N(1,1)$, then $X \leq_{st} Y$. To show this, note that $(\widehat{X}, \widehat{Y}) = (X, 1 + X)$ is a coupling of (X, Y). Because $\widehat{X} \leq \widehat{Y}$, we must have $\widehat{X} \leq_{st} \widehat{Y}$ and thus $X \leq_{st} Y$.

   Even though the probability that two independent continuous random
variables exactly equal each other is always zero, it is possible to couple two
variables with completely different density functions so that they equal each
other with high probability. For the random variables (X, Y), the coupling
$(\widehat{X}, \widehat{Y})$ is called a *maximal coupling* if $P(\widehat{X} = \widehat{Y})$ is as large as possible. We
next show how large this probability can be and how to create a maximal
coupling.

Proposition 2.5 *Suppose X and Y are random variables with respective
piecewise continuous density functions f and g. The maximal coupling
$(\widehat{X}, \widehat{Y})$ for (X, Y) has*

$$P(\widehat{X} = \widehat{Y}) = \int_{-\infty}^{\infty} \min(f(x), g(x))dx.$$

Proof Letting $p = \int_{-\infty}^{\infty} \min(f(x), g(x))dx$ and $A = \{x : f(x) < g(x)\}$, note
that any coupling $(\widehat{X}, \widehat{Y})$ of (X, Y) must satisfy

$$
\begin{aligned}
P(\widehat{X} = \widehat{Y}) &= P(\widehat{X} = \widehat{Y} \in A) + P(\widehat{X} = \widehat{Y} \in A^c) \\
&\le P(X \in A) + P(Y \in A^c) \\
&= \int_A f(x)dx + \int_{A^c} g(x)dx \\
&= p.
\end{aligned}
$$

We use the fact that f, g are piecewise continuous to justify that the integrals in the preceding step (and in the next step) are well defined. Next
we construct a coupling with $P(\widehat{X} = \widehat{Y}) \ge p$, which in light of the previous
inequality, must therefore be the maximal coupling. Let B, C, and D be
independent random variables with respective density functions

$$b(x) = \frac{\min(f(x), g(x))}{p}$$

$$c(x) = \frac{f(x) - \min(f(x), g(x))}{1 - p}$$

and

$$d(x) = \frac{g(x) - \min(f(x), g(x))}{1 - p}.$$

Let I be a Bernoulli(p) random variable independent of B, C, and D, and
if $I = 1$, then let $\widehat{X} = \widehat{Y} = B$, and otherwise, let $\widehat{X} = C$ and $\widehat{Y} = D$. This
clearly gives $P(\widehat{X} = \widehat{Y}) \ge P(I = 1) = p$ and

$$P(\widehat{X} \le x) = P(\widehat{X} \le x | I = 1)p + P(\widehat{X} \le x | I = 0)(1 - p)$$

$$= p \int_{-\infty}^{x} b(x)dx + (1 - p) \int_{-\infty}^{x} c(x)dx$$

$$= \int_{-\infty}^{x} f(x)dx,$$

and the same argument again gives $P(\widehat{Y} \le x) = P(Y \le x)$. ∎

We can repeat this exact same proof with probability mass functions replacing densities and sums replacing integrals to obtain the following proposition for two discrete random variables.

Proposition 2.6 *Suppose X and Y are discrete random variables, with each taking values in a set A. Let their respective probability mass functions $f(x) = P(X = x)$ and $g(x) = P(Y = x)$. The maximal coupling of (X, Y) has*

$$P(\widehat{X} = \widehat{Y}) = \sum_{x} \min \left(g(x), f(x) \right).$$

There is a relationship between how closely two variables can be coupled and how close they are in distribution. One common measure of distance between the distributions of two random variables X and Y is called *total variation distance*, which is defined as

$$d_{TV}(X, Y) = \sup_{A} |P(X \in A) - P(Y \in A)|.$$

We next show the link between total variation distance and couplings.

Proposition 2.7 *If $(\widehat{X}, \widehat{Y})$ is a maximal coupling for (X, Y), then*

$$d_{TV}(X, Y) = P(\widehat{X} \ne \widehat{Y}).$$

Proof The result will be proven under the assumption that X, Y are continuous with respective density functions f, g. Letting $A = \{x : f(x) > g(x)\}$, we must have

$d_{TV}(X, Y)$
$$= \max\{P(X \in A) - P(Y \in A), P(Y \in A^c) - P(X \in A^c)\}$$
$$= \max\{P(X \in A) - P(Y \in A), 1 - P(Y \in A) - 1 + P(X \in A)\}$$
$$= P(X \in A) - P(Y \in A)$$

and so

$$P(\widehat{X} \neq \widehat{Y}) = 1 - \int_{-\infty}^{\infty} \min(f(x), g(x)) dx$$
$$= 1 - \int_A g(x) dx - \int_{A^c} f(x) dx$$
$$= 1 - P(Y \in A) - 1 + P(X \in A)$$
$$= d_{TV}(X, Y).$$

∎

2.3   Poisson Approximation and Le Cam's Theorem

It's well known that a binomial distribution converges to a Poisson distribution when the number of trials is increased and the probability of success is decreased at the same time in such a way that the mean stays constant. This also motivates using a Poisson distribution as an approximation for a binomial distribution if the probability of success is small and the number of trials is large. If the number of trials is very large, computing the distribution function of a binomial distribution can be computationally difficult, whereas the Poisson approximation may be much easier.

A fact that is not as well known is that the Poisson distribution can be a reasonable approximation even if the trials have varying probabilities and even if the trials are not completely independent of each other. This approximation is interesting because dependent trials are notoriously difficult to analyze in general, and the Poisson distribution is elementary.

It's possible to assess how accurate such Poisson approximations are, and we first give a bound on the error of the Poisson approximation for completely independent trials with different probabilities of success.

Proposition 2.8 *Let X_i be independent Bernoulli(p_i), and let $W = \sum_{i=1}^{n} X_i$, $Z \sim Poisson(\lambda)$, and $\lambda = E[W] = \sum_{i=1}^{n} p_i$. Then*

$$d_{TV}(W, Z) \leq \sum_{i=1}^{n} p_i^2.$$

Proof We first write $Z = \sum_{i=1}^{n} Z_i$, where the Z_i are independent and $Z_i \sim$ Poisson(p_i). Then we create the maximal coupling $(\widehat{Z}_i, \widehat{X}_i)$ of (Z_i, X_i) and use the previous corollary to get

$$P(\widehat{Z}_i = \widehat{X}_i) = \sum_{k=0}^{\infty} \min(P(X_i = k), P(Z_i = k))$$
$$= \min(1 - p_i, e^{-p_i}) + \min(p_i, p_i e^{-p_i})$$
$$= 1 - p_i + p_i e^{-p_i}$$
$$\geq 1 - p_i^2,$$

where we use $e^{-x} \geq 1 - x$. Using that $(\sum_{i=1}^{n} \widehat{Z}_i, \sum_{i=1}^{n} \widehat{X}_i)$ is a coupling of (Z, W) yields that

$$d_{TV}(W, Z) \leq P\left(\sum_{i=1}^{n} \widehat{Z}_i \neq \sum_{i=1}^{n} \widehat{X}_i\right)$$
$$\leq P(\cup_i \{\widehat{Z}_i \neq \widehat{X}_i\})$$
$$\leq \sum_i P(\widehat{Z}_i \neq \widehat{X}_i)$$
$$\leq \sum_{i=1}^{n} p_i^2.$$

■

Remark 2.9 The drawback to this beautiful result is that when $E[W]$ is large the upper bound could be much greater than 1 even if W has approximately a Poisson distribution. For example, if W is a binomial (100, 0.1) random variable and $Z \sim$ Poisson(10), we should have $W \approx_d Z$, but the proposition gives us the trivial and useless result $d_{TV}(W, Z) \leq 100 \times (0.1)^2 = 1$.

2.4   Stein–Chen Method

The Stein–Chen method is another approach for obtaining an upper bound on $d_{TV}(W, Z)$, where Z is a Poisson random variable and W another variable of interest. This approach covers the distribution of the number of successes in both dependent and independent trials with varying probabilities of success.

In order for the bound to be good, it should be close to 0 when W is close in distribution to Z. In order to achieve this, the Stein–Chen method uses the interesting property that if Z is a Poisson random variable with mean λ then

$$kP(Z = k) = \lambda P(Z = k - 1). \tag{2.1}$$

Thus, for any bounded function f with $f(0) = 0$,

$$
\begin{aligned}
E[Zf(Z)] &= \sum_{k=0}^{\infty} kf(k)P(Z = k) \\
&= \lambda \sum_{k=1}^{\infty} f(k)P(Z = k - 1) \\
&= \lambda \sum_{i=0}^{\infty} f(i+1)P(Z = i) \\
&= \lambda E[f(Z+1)].
\end{aligned}
$$

The secret to using the preceding is in cleverly picking a function f such that $d_{TV}(W, Z) \leq E[Wf(W)] - \lambda E[f(W+1)]$, and so we are likely to get a small upper bound when $W \approx_d Z$, where we use the notation $W \approx_d Z$ to mean that W and Z have approximately the same distribution.

Suppose $Z \sim \text{Poisson}(\lambda)$ and A is any set of nonnegative integers. Define the function $f_A(k), k = 0, 1, 2, ...$, starting with $f_A(0) = 0$ and then using the following Stein equation for the Poisson distribution:

$$
\lambda f_A(k+1) - kf_A(k) = I_{k \in A} - P(Z \in A). \tag{2.2}
$$

Notice that by plugging in any random variable W for k and taking expected values we get

$$
\lambda E[f_A(W+1) - E[Wf_A(W)]] = P(W \in A) - P(Z \in A),
$$

so that

$$
d_{TV}(W, Z) = \sup_A |\lambda E[f_A(W+1)] - E[Wf_A(W)]|. \tag{2.3}
$$

Lemma 2.10 *For any A and i, j, $|f_A(i) - f_A(j)| \leq \min(1, 1/\lambda)|i - j|$.*

Proof The solution to Equation 2.2 is

$$
f_A(k+1) = \sum_{j \in A} \frac{I_{j \leq k} - P(Z \leq k)}{\lambda P(Z = k)/P(Z = j)}
$$

because when we plug it in to the left-hand side of Equation 2.2 and use Equation 2.1 in the second line in the following, we get

$$
\begin{aligned}
&\lambda f_A(k+1) - kf_A(k) \\
&= \sum_{j \in A} \left(\frac{I_{j \leq k} - P(Z \leq k)}{P(Z = k)/P(Z = j)} - k \frac{I_{j \leq k-1} - P(Z \leq k-1)}{\lambda P(Z = k-1)/P(Z = j)} \right) \\
&= \sum_{j \in A} \frac{I_{j=k} - P(Z = k)}{P(Z = k)/P(Z = j)} \\
&= I_{k \in A} - P(Z \in A),
\end{aligned}
$$

which is the right-hand side of Equation 2.2.

Because

$$\frac{P(Z \leq k)}{P(Z = k)} = \sum_{i \leq k} \frac{k! \lambda^{i-k}}{i!} = \sum_{i \leq k} \frac{k! \lambda^{-i}}{(k-i)!}$$

is increasing in k and

$$\frac{1 - P(Z \leq k)}{P(Z = k)} = \sum_{i > k} \frac{k! \lambda^{i-k}}{i!} = \sum_{i > 0} \frac{k! \lambda^{i}}{(i+k)!}$$

is decreasing in k, we see that $f_{\{j\}}(k+1) \leq f_{\{j\}}(k)$ when $j \neq k$ and thus

$$
\begin{aligned}
f_A(k+1) - f_A(k) &= \sum_{j \in A} f_{\{j\}}(k+1) - f_{\{j\}}(k) \\
&\leq \sum_{j \in A, j=k} f_{\{j\}}(k+1) - f_{\{j\}}(k) \\
&\leq \frac{P(Z > k)}{\lambda} + \frac{P(Z \leq k-1)}{\lambda P(Z = k-1)/P(Z = k)} \\
&= \frac{P(Z > k)}{\lambda} + \frac{P(Z \leq k-1)}{k} \\
&\leq \frac{P(Z > k)}{\lambda} + \frac{P(0 < Z \leq k)}{\lambda} \\
&\leq \frac{1 - e^{-\lambda}}{\lambda} \\
&\leq \min(1, 1/\lambda)
\end{aligned}
$$

and

$$-f_A(k+1) + f_A(k) = f_{A^c}(k+1) - f_{A^c}(k) \leq \min(1, 1/\lambda),$$

which together give

$$|f_A(k+1) - f_A(k)| \leq \min(1, 1/\lambda).$$

The final result is proved using

$$|f_A(i) - f_A(j)| \leq \sum_{k=\min(i,j)}^{\max(i,j)-1} |f_A(k+1) - f_A(k)| \leq |j - i| \min(1, 1/\lambda).$$

∎

Theorem 2.11 *Suppose* $W = \sum_{i=1}^{n} X_i$, *where* X_i *are indicator variables with* $P(X_i = 1) = \lambda_i$ *and* $\lambda = \sum_{i=1}^{n} \lambda_i$. *Letting* $Z \sim$ *Poisson*(λ) *and* $V_i =_d (W - 1|X_i = 1)$, *we have*

$$d_{TV}(W, Z) \leq \min(1, 1/\lambda) \sum_{i=1}^{n} \lambda_i E|W - V_i|.$$

Proof We use

$$E[Wf_A(W)] = \sum_{i=1}^{n} E[X_i f_A(W)]$$

$$= \sum_{i=1}^{n} E[X_i f_A(W)|X_i = 1]\lambda_i$$

$$= \sum_{i=1}^{n} E[f_A(V_i + 1)]\lambda_i$$

with Equation 2.3 to get

$$d_{TV}(W, Z) = \sup_{A} |E[\lambda E[f_A(W + 1)] - Wf_A(W)]|$$

$$\leq \sup_{A} \sum_{i=1}^{n} \lambda_i |E[f_A(W + 1) - f_A(V_i + 1)]|$$

$$\leq \sup_{A} \sum_{i=1}^{n} \lambda_i E|f_A(W + 1) - f_A(V_i + 1)|$$

$$\leq \min(1, 1/\lambda) \sum_{i=1}^{n} \lambda_i E|W - V_i|,$$

∎

Proposition 2.12 *With the preceding notation, we have the following:*

1. *If $W \geq V_i$ almost surely for all i, then we have*

$$d_{TV}(W, Z) \leq 1 - \text{Var}(W)/E[W].$$

2. *If either $W \geq V_i$ almost surely or $W \leq V_i$ almost surely for all i, then*

$$d_{TV}(W, Z) \leq \min(1, 1/\lambda) \sum_{i=1}^{n} \lambda_i |E[W] - E[V_i]|.$$

Proof Using $W \geq V_i$, we have

$$\sum_{i=1}^{n} \lambda_i E|W - V_i| = \sum_{i=1}^{n} \lambda_i E[W - V_i]$$

$$= \lambda^2 + \lambda - \sum_{i=1}^{n} \lambda_i E[1 + V_i]$$

$$= \lambda^2 + \lambda - \sum_{i=1}^{n} \lambda_i E[W | X_i = 1]$$

$$= \lambda^2 + \lambda - \sum_{i=1}^{n} E[X_i W]$$

$$= \lambda - \mathrm{Var}(W),$$

and the preceding theorem along with $\lambda = E[W]$ gives

$$d_{TV}(W, Z) \leq \min(1, 1/E[W])(E[W] - \mathrm{Var}(W))$$
$$\leq 1 - \mathrm{Var}(W)/E[W]$$

proving part 1. Under condition 2, we will have $E|W - V_i| = |E[W - V_i]|$, so the result also follows from the previous theorem. ∎

Example 2.13 Let X_i be independent Bernoulli(p_i) random variables with $\lambda = \sum_{i=1}^{n} p_i$. Let $W = \sum_{i=1}^{n} X_i$, and let $Z \sim \mathrm{Poisson}(\lambda)$. Using $V_i = \sum_{j \neq i} X_j$, note that $W \geq V_i$ and $E[W - V_i] = p_i$, so the preceding theorem gives us

$$d_{TV}(W, Z) \leq \min(1, 1/\lambda) \sum_{i=1}^{n} p_i^2.$$

For instance, if X is a binomial random variable with parameters $n = 100, p = 1/10$, then the upper bound on the total variation distance between X and a Poisson random variable with mean 10 given by the Stein–Chen method is $1/10$, as opposed to the upper bound of 1 that results from the LeCam method of the preceding section.

Example 2.14 A coin with probability $p = 1 - q$ of coming up heads is flipped $n + k$ times. We are interested in $P(R_k)$, where R_k is the event that a run of at least k heads in a row occurs. To approximate this probability, the exact expression for which is given in Example 1.10, let $X_i = 1$ if flip i lands tails and flips $i + 1, \ldots, i + k$ all land heads; otherwise, let $X_i = 0$, $i = 1, \ldots, n$. Let

$$W = \sum_{i=1}^{n} X_i$$

and note that there will be a run of k heads either if $W > 0$ or if the first k flips all land heads. Consequently,

$$P(W > 0) < P(R_k) < P(W > 0) + p^k.$$

Because the flips are independent and $X_i = 1$ implies $X_j = 0$ for all $j \neq i$ where $|i - j| \leq k$, it follows that if we let

$$V_i = W - \sum_{j=i-k}^{i+k} X_j \qquad ,$$

then $V_i =_d (W - 1|X_i = 1)$ and $W \geq V_i$. Using $\lambda = E[W] = nqp^k$ and $E[W - V_i] = (2k + 1)qp^k$, we see that

$$d_{TV}(W, Z) \leq \min(1, 1/\lambda)n(2k + 1)q^2p^{2k},$$

where $Z \sim \text{Poisson}(\lambda)$. For instance, suppose we flip a fair coin 1,034 times and want to approximate the probability that we have a run of 10 heads in a row. In this case, $n = 1,024, k = 10$, and $p = 1/2$, so $\lambda = 1,024/2^{11} = 0.5$. Consequently,

$$P(W > 0) \approx 1 - e^{-0.5}$$

with the error in the preceding approximation being at most $21/2^{12}$. Consequently, we obtain

$$1 - e^{-0.5} - 21/2^{12} < P(R_{10} > 0) < 1 - e^{-0.5} + 21/2^{12} + (1/2)^{10}$$

or

$$0.388 < P(R_{10} > 0) < 0.4.$$

Example 2.15 *Birthday Problem.* With m people and n days in the year, let Y_i equal the number of people born on day i. Let $X_i = I_{Y_i=0}$, and $W = \sum_{i=1}^{n} X_i$ equal the number of days on which nobody has a birthday.

Next imagine n different hypothetical scenarios are constructed, where in scenario i all the Y_i people initially born on day i have their birthdays reassigned randomly to other days. Let $1 + V_i$ equal the number of days under scenario i on which nobody has a birthday, and note that $V_i =_d (W - 1|X_i = 1)$.

Notice that this construction gives $W \geq V_i$, so $E|W - V_i| = |E[W] - E[V_i]|$. Letting $\lambda = E[W] = n(1 - 1/n)^m$ and noting $E[V_i] = (n - 1)(1 - 1/(n - 1))^m$, we use Theorem 2.11 with $Z \sim \text{Poisson}(\lambda)$ to get

$$d_{TV}(W, Z) \leq \min(1, \lambda)(\lambda - (n - 1)(1 - 1/(n - 1))^m).$$

2.5 Stein's Method for the Geometric Distribution

In this section, we show how to obtain an upper bound on the distance $d_{TV}(W, Z)$, where W is a given random variable and Z has a geometric distribution with parameter $p = 1 - q = P(W = 1) = P(Z = 1)$. We use a version of Stein's method applied to the geometric distribution. Define $f_A(1) = 0$, and for $k = 1, 2, \ldots$, use the recursion

$$f_A(k) - q f_A(k + 1) = I_{k \in A} - P(Z \in A).$$

Lemma 2.16 *We have* $|f_A(i) - f_A(j)| < 1/p$

Proof It's easy to check that the solution is

$$f_A(k) = P(Z \in A, Z \geq k)/P(Z = k) - P(Z \in A)/p,$$

and because neither of the two terms in the difference can be larger than $1/p$, the lemma follows. ∎

Theorem 2.17 *Given random variables W and V such that $V =_d (W - 1|$ $W > 1)$, let $p = P(W = 1)$ and $Z \sim geometric(p)$. Then*

$$d_{TV}(W, Z) \leq q p^{-1} P(W \neq V).$$

Proof

$$
\begin{aligned}
|P(W \in A) - P(Z \in A)| &= |E[f_A(W) - q f_A(W + 1)]| \\
&= |q E[f_A(W)|W > 1] - q E[f_A(W + 1)]| \\
&\leq q E|f_A(1 + V) - f_A(1 + W)| \\
&\leq q p^{-1} P(W \neq V),
\end{aligned}
$$

where the last inequality follows from the preceding lemma. ∎

Example 2.18 *Coin flipping.* A coin has probability $h = 1 - q$ of coming up heads with each flip, and let $X_i = H$ if the ith flip is heads and $X_i = T$ if it is tails. We are interested in the distribution of the number of flips

$$M = \min\{i \geq 1 : (X_i, X_{i+1}, \ldots, X_{i+k-1}) = (H, H, \ldots, H)\}$$

required until the start of the first appearance of a run of k heads in a row. Suppose in particular we are interested in estimating $P(M \in A)$ for some set A. To do this, define the number of flips

$$N = \min\{i \geq 1 : (X_i, X_{i+1}, \ldots, X_{i+k}) = (T, H, \ldots, H)\}$$

required until the start of the first appearance of a run of a tail followed by k heads in a row. For instance, if $k = 3$ and the flip sequence is $HHTTHHH$, then $N = 4$ and $M = 5$. Note that we will have $N + 1 = M$ unless $M = 1$ so

$$d_{TV}(M, N+1) \leq P(N+1 \neq M) = P(M = 1) = p^k.$$

We will first obtain a bound on the geometric approximation to N and use it to get a bound for M. We first define

$$W = \min\{i \geq 2 : (X_i, X_{i+1}, \ldots, X_{i+k}) = (T, H, \ldots, H)\} - 1$$

and note that $W =_d N$. Then, we define the event

$$A = \{(X_1, X_2, \ldots, X_{k+1}) = (T, H, \ldots, H)\}$$

and independently generate new variables that have the joint distribution

$$(X'_1, X'_2, \ldots, X'_{k+1}) =_d (X_1, X_2, \ldots, X_{k+1} | A^c)$$

given $(T, H, \ldots, H)$ doesn't appear first. We will construct a new sequence of coin flips $Y_1, Y_2, \ldots$ by letting

$$Y_i = \begin{cases} X'_i & \text{if } i \leq k+1 \text{ and } A \text{ happens} \\ X_i & \text{otherwise} \end{cases}$$

and note that with

$$V = \min\{i \geq 2 : (Y_i, Y_{i+1}, \ldots, Y_{i+k}) = (T, H, \ldots, H)\} - 1$$

we have $V =_d (W - 1 | W > 1)$ and

$$P(W \neq V) \leq P(A \cap \{V \leq k\}) \leq \frac{kqp^{2k}}{1 - qp^k}.$$

Thus, if $Z \sim \text{geometric}(qp^k)$, the previous theorem gives

$$d_{TV}(N, Z) \leq \frac{1 - qp^k}{qp^k} P(W \neq V) \leq kp^k$$

and the triangle inequality applies to d_{TV} so we get

$$d_{TV}(M - 1, Z) \leq d_{TV}(M - 1, N) + d_{TV}(N, Z) = (k+1)p^k,$$

and we get the bounds

$$P(Z \in A) - (k+1)p^k \leq P(M \in A) \leq P(Z \in A) + (k+1)p^k.$$

2.6 Stein's Method for the Normal Distribution

Let $Z \sim N(0,1)$ be a standard normal random variable. It can be shown that for smooth functions f we have $E[f'(Z) - Zf(Z)] = 0$, and this inspires the following Stein equation.

Lemma 2.19 *Given $\alpha > 0$ and any value of z let*

$$h_{\alpha,z}(x) \equiv h(x) = \begin{cases} 1 & \text{if } x \leq z \\ 0 & \text{if } x \geq z + \alpha \\ (\alpha + z - x)/\alpha & \text{otherwise,} \end{cases}$$

and define the function $f_{\alpha,z}(x) \equiv f(x), -\infty < x < \infty$ so it satisfies

$$f'(x) - xf(x) = h(x) - E[h(Z)].$$

Then $|f'(x) - f'(y)| \leq \frac{2}{\alpha}|x - y|, \forall x, y.$

Proof Letting $\phi(x) = e^{-x^2/2}/\sqrt{2\pi}$ be the standard normal density function, we have the solution

$$f(x) = \frac{E[h(Z)I_{Z \leq x}] - E[h(Z)]P(Z \leq x)}{\phi(x)},$$

which can be checked by differentiating using $\frac{d}{dx}(\phi(x)^{-1}) = x/\phi(x)$ to get $f'(x) = xf(x) + h(x) - E[h(Z)]$. Then this gives

$$|f''(x)| = |f(x) + xf'(x) + h'(x)|$$
$$= |(1 + x^2)f(x) + x(h(x) - E[h(Z)]) + h'(x)|.$$

Because

$$h(x) - E[h(Z)]$$
$$= \int_{-\infty}^{\infty} (h(x) - h(s))\phi(s)ds$$
$$= \int_{-\infty}^{x} \int_{s}^{x} h'(t)dt\phi(s)ds - \int_{x}^{\infty} \int_{x}^{s} h'(t)dt\phi(s)ds$$
$$= \int_{-\infty}^{x} h'(t)P(Z \leq t)dt - \int_{x}^{\infty} h'(t)P(Z > t))dt,$$

and a similar argument gives

$$f(x) = -\frac{P(Z > x)}{\phi(x)} \int_{-\infty}^{x} h'(t)P(Z \leq t)dt$$
$$- \frac{P(Z \leq x)}{\phi(x)} \int_{x}^{\infty} h'(t)P(Z > t)dt,$$

we get

$$|f''(x)| \le |h'(x)| + |(1+x^2)f(x) + x(h(x) - E[h(Z)]|$$

$$\le \frac{1}{\alpha}$$

$$+ \frac{1}{\alpha}(-x + (1+x^2)P(Z > x)/\phi(x))(xP(Z \le x) + \phi(x))$$

$$+ \frac{1}{\alpha}(x + (1+x^2)P(Z \le x)/\phi(x))(-xP(Z > x) + \phi(x))$$

$$\le 2/\alpha.$$

We finally use

$$|f'(x) - f'(y)| \le \int_{\min(x,y)}^{\max(x,y)} |f''(x)|dx \le \frac{2}{\alpha}|x - y|$$

to give the lemma. ∎

Theorem 2.20 *If $Z \sim N(0,1)$ and $W = \sum_{i=1}^{n} X_i$, where X_i are independent variables with mean 0 and $\mathrm{Var}(W) = 1$, then*

$$\sup_z |P(W \le z) - P(Z \le z)| \le 2\sqrt{3\sum_{i=1}^{n} E[|X_i|^3]}.$$

Proof Given any $\alpha > 0$ and any z, define h, f as in the previous lemma. Then

$$P(W \le z) - P(Z \le z)$$
$$= Eh(W) - Eh(Z) + Eh(Z) - P(Z \le z)$$
$$\le |E[h(W)] - E[h(Z)]| + P(z \le Z \le z + \alpha)$$
$$= |E[h(W)] - E[h(Z)]| + \int_z^{z+\alpha} \frac{1}{\sqrt{2\pi}}e^{-x^2/2}dx$$
$$\le |E[h(W)] - E[h(Z)]| + \alpha.$$

To finish the proof of the theorem, we show

$$|E[h(W)] - E[h(Z)]| \le \sum_{i=1}^{n} 3E[|X_i^3|]/\alpha,$$

and then by choosing

$$\alpha = \sqrt{\sum_{i=1}^{n} 3E[|X_i^3|]}$$

we get

$$P(W \leq x) - P(Z \leq x) \leq 2\sqrt{3 \sum_{i=1}^{n} E[|X_i|^3]}.$$

Repeating the same argument starting with

$$\begin{aligned}
P(Z \leq z) &- P(W \leq z) \\
&\leq P(Z \leq z) - Eh(Z + \alpha) + Eh(Z + \alpha) - Eh(W + \alpha) \\
&\leq |E[h(W + \alpha)] - E[h(Z + \alpha)]| + P(z \leq Z \leq z + \alpha),
\end{aligned}$$

the theorem is proved.

To do this, let $W_i = W - X_i$, and let Y_i be a random variable independent of all else that has the same distribution as X_i. Using $\text{Var}(W) = \sum_{i=1}^{n} E[Y_i^2] = 1$ and $E[X_i f(W_i)] = E[X_i] E[f(W_i)] = 0$ in the second equality to come, and the preceding lemma with $|W - W_i - t| \leq |t| + |X_i|$ in the second inequality to come, we have

$$\begin{aligned}
|E[h(W)] &- E[h(Z)]| \\
&= |E[f'(W) - W f(W)]| \\
&= \left| \sum_{i=1}^{n} E[Y_i^2 f'(W) - X_i(f(W) - f(W_i))] \right| \\
&= \left| \sum_{i=1}^{n} E\left[Y_i \int_0^{Y_i} (f'(W) - f'(W_i + t)) dt \right] \right| \\
&\leq \sum_{i=1}^{n} E\left[Y_i \int_0^{Y_i} |f'(W) - f'(W_i + t)| dt \right] \\
&\leq \sum_{i=1}^{n} E\left[Y_i \int_0^{Y_i} \frac{2}{\alpha} (|t| + |X_i|) dt \right],
\end{aligned}$$

and continuing on from this we get

$$\begin{aligned}
&= \sum_{i=1}^{n} E[|X_i^3|]/\alpha + 2E[X_i^2] E[|X_i|]/\alpha \\
&\leq \sum_{i=1}^{n} 3E[|X_i^3|]/\alpha,
\end{aligned}$$

where in the last line we use $E[|X_i|] E[X_i^2] \leq E[|X_i|^3]$ (which follows because $|X_i|$ and X_i^2 are both increasing functions of $|X_i|$ and are thus positively correlated. The proof of this result is given in the following lemma ∎

Lemma 2.21 *If $f(x)$ and $g(x)$ are nondecreasing functions, then for any random variable X*

$$E[f(X)g(X)] \geq E[f(X)]E[g(X)].$$

Proof Let X_1 and X_2 be independent random variables having the same distribution as X. Because $f(X_1) - f(X_2)$ and $g(X_1) - g(X_2)$ are either both nonnegative or are both nonpositive, their product is nonnegative. Consequently,

$$E[(f(X_1) - f(X_2))(g(X_1) - g(X_2))] \geq 0,$$

or equivalently,

$$E[f(X_1)g(X_1)] + E[f(X_2)g(X_2)] \geq E[f(X_1)g(X_2)] + E[f(X_2)g(X_1)].$$

But by independence,

$$E[f(X_1)g(X_2)] = E[f(X_2)g(X_1)] = E[f(X)]E[g(X)],$$

and the result follows. ■

The preceding results yield the following version of the central limit theorem as an immediate corollary.

Corollary 2.22 *If $Z \sim N(0,1)$ and $Y_1, Y_2, \ldots$ are iid random variables with $E[Y_i] = \mu$, $\mathrm{Var}(Y_i) = \sigma^2$ and $E[|Y_i|^3] < \infty$, then as $n \to \infty$ we have*

$$P\left(\frac{1}{\sqrt{n}} \sum_{i=1}^{n} \frac{Y_i - \mu}{\sigma} \leq z\right) \to P(Z \leq z).$$

Proof Letting $X_i = \frac{Y_i - \mu}{\sigma\sqrt{n}}, i \geq 1$, and $W_n = \sum_{i=1}^{n} X_i$, then W_n satisfies the conditions of Theorem 2.20. Because

$$\sum_{i=1}^{n} E[|X_i|^3] = nE[|X_1|^3] = \frac{nE[|Y_1 - \mu|^3]}{\sigma^3 n^{3/2}} \to 0,$$

it follows from Theorem 2.20 that $P(W_n \leq x) \to P(Z \leq x)$. ■

See [3] for additional results on applying Stein's method to normal distributions.

2.7 Stein's Method for the Exponential Distribution

Let Z have an exponential distribution with mean 1. Here we develop Stein's method and use it to approximate the sum of a geometric number of independent random variables.

Lemma 2.23 *Given $\alpha > 0$ and $z \geq 0$, let $h(x)$ be defined as in Lemma 2.19, and define the function $f_{\alpha,z}(x) \equiv f(x), 0 \leq x < \infty$, so it satisfies $f(0) = 0$ and, for $x > 0$,*

$$f'(x) - f(x) = h(x) - E[h(Z)].$$

Then $|f'(x) - f'(y)| \leq \frac{2}{\alpha}|x - y|, \forall x, y \geq 0$.

Proof We have the solution

$$f'(x) = -e^x \int_x^\infty h'(t)e^{-t}dt,$$

which can be checked by taking the derivative of both sides to get $f''(x) = h'(x) + f'(x)$. Noting that $|h'(x)| \leq 1/\alpha$, we have

$$|f'(x)| \leq e^x \int_x^\infty \frac{e^{-t}}{\alpha}dt \leq 1/\alpha$$

and

$$|f''(x)| = |h'(x) + f'(x)| \leq 2/\alpha.$$

∎

We can use this to obtain the following result for a geometric sum.

Corollary 2.24 *For $0 < p < 1$ let $N \sim Geometric(p)$, suppose $X_i \geq 0$ are iid and independent of N with $E[X_i] = p$, let $S_n = \sum_{i=1}^n X_i$, and let $W = S_N$. With $Z \sim Exponential(1)$, we have*

$$\sup_z |P(W \leq z) - P(Z \leq z)| \leq \sqrt{4E[X_1^2]/p + 8p}.$$

Before giving the proof, for a random variable $X \geq 0$, we say that X^s is a size-biased version of X if $E[f(X^s)] = E[Xf(X)]/E[X]$ for all bounded functions f. For example, if X is discrete, then $P(X^s = k) = kP(X = k)/E[X]$, and if X is continuous with density $g(x)$, then X^s has density $xg(x)/E[X]$.

Proof of Corollary Let $U \sim U(0,1)$ be independent of all else and let $W^* = UX_N^s + S_{N-1}$, where X_N^s is a size-biased version of X_N and is independent of U and S_{N-1}. Given any $\alpha > 0$ and $z \geq 0$, define h, f as in the previous lemma. Then

$$
\begin{aligned}
P(W \leq z) &- P(Z \leq z) \\
&= Eh(W) - Eh(Z) + Eh(Z) - P(Z \leq z) \\
&\leq |E[h(W)] - E[h(Z)]| + P(z \leq Z \leq z + \alpha) \\
&= |E[f'(W) - f(W)]| + \int_z^{z+\alpha} e^{-x}dx \\
&\leq |E[f'(W) - f'(W^*)]| + \alpha \\
&\leq 2E|W - W^*|/\alpha + \alpha,
\end{aligned}
$$

where we use the previous lemma in the last line and

$$
\begin{aligned}
E[f'(W^*)] &= E[X_N f'(UX_N + S_{N-1})]/E[X_N] \\
&= E\left[\int_0^1 X_N f'(uX_N + S_{N-1})du\right]/p \\
&= E[f(S_N) - f(S_{N-1})]/p \\
&= \sum_{n=1}^{\infty} P(N = n)E[f(S_n) - f(S_{n-1})]/p \\
&= \sum_{n=1}^{\infty} P(N \geq n)E[f(S_n) - f(S_{n-1})] \\
&= E\left[\sum_{n=1}^{\infty} 1_{N \geq n}(f(S_n) - f(S_{n-1}))\right] \\
&= E[f(W)]
\end{aligned}
$$

in the fourth line. Then, choosing $\alpha = \sqrt{2E|W - W^*|}$, we obtain the result after noting

$$
E|W - W^*| \leq E[UX_N^s] + E[X_N] = E[X_1^2]/(2p) + p.
$$

■

2.8 Exercises

1. If $X \sim \text{Poisson}(a)$ and $Y \sim \text{Poisson}(b)$, with $b > a$, use coupling to show that $Y \geq_{st} X$.

2. (a) Show that $X \leq_{st} Y$ if and only if $E[h(X)] \leq E[h(Y)]$ for all increasing functions h.

   (b) Suppose $X_1, \ldots, X_n$ are independent, and $Y_1, \ldots, Y_n$ are independent. Show that if $X_i \leq_{st} Y_i$ for all $i = 1, \ldots, n$ then $E[h(X_1, \ldots, X_n)] \leq E[h(Y_1, \ldots, Y_n)]$ for all increasing functions h.

3. Suppose two particles start at position 0 and at each time period particle i moves one position to the right with probability $p_{i,j}$ or moves one position to the left with probability $1 - p_{i,j}$ where j is the position of the particle before it moves. Let $X_{n,i}$ be the position of particle i after n moves. If $p_{2,j} \geq p_{1,j}$ for all j show that $X_{n,2} \geq_{st} X_{n,1}$ for all n. Is this also always true under the same conditions but allowing the second particle to initially start to the right of the first particle?

4. Let X, Y be indicator variables with $E[X] = a$ and $E[Y] = b$.

   (a) Show how to construct a maximal coupling $\widehat{X}, \widehat{Y}$ for X and Y, and then compute $P(\widehat{X} = \widehat{Y})$ as a function of a, b.

   (b) Show how to construct a minimal coupling to minimize $P(\widehat{X} = \widehat{Y})$.

5. In a room full of n people, let X be the number of people who share a birthday with at least one other person in the room. Then let Y be the number of pairs of people in the room having the same birthday. Assume that the n birthdays are independent and that each is equally likely to be any of the 365 days of the year.

   (a) Compute $E[X]$ and $\text{Var}(X)$ and $E[Y]$ and $\text{Var}(Y)$.

   (b) Which of the two variables X or Y do you believe will more closely follow a Poisson distribution? Why?

   (c) If $n = 51$, use a Poisson approximation to estimate $P(X > 9)$ and $P(Y > 6)$. Which of these two approximations do you think will be better? Have we observed a rare event here?

6. Compute a bound on the accuracy of the better approximation in part (c) of Exercise 5 using the Stein–Chen method.

7. For discrete X, Y prove $d_{TV}(X, Y) = \frac{1}{2} \sum_x |P(X = x) - P(Y = x)|$

8. For discrete X, Y show that $P(X \neq Y) \geq d_{TV}(X, Y)$ and that a coupling exists that yields equality.

9. Compute a bound on the accuracy of a normal approximation for a Poisson random variable with mean 100.

10. If $X \sim \text{Geometric}(p)$, with $q = 1 - p$, then show that for any bounded function f with $f(1) = 0$, we have $E[f(X) - qf(X + 1)] = 0$.

11. Suppose $X_1, X_2, \ldots$ are independent mean zero random variables with $|X_n| < 1$ for all n and $\sum_{i \leq n} \text{Var}(X_i)/n \to s < \infty$. If $S_n = \sum_{i \leq n} X_i$ and $Z \sim N(0, s)$, show that $S_n/\sqrt{n} \to_p Z$.

12. Suppose m balls are placed among n urns, with each ball independently going in to urn i with probability p_i. Assume m is much larger than n. Approximate the chance none of the urns are empty, and give a bound on the error of the approximation.

13. A ring with a circumference of c is cut into n pieces (where n is large) by cutting at n places chosen uniformly at random around the ring. Estimate the chance you get k pieces of length at least a, and give a bound on the error of the approximation.

14. Suppose X_i, $i = 1, 2, ..., 10$ are iid $U(0,1)$. Give an approximation for $P(\sum_{i=1}^{10} X_i > 7)$, and give a bound on the error of this approximation.

15. Suppose X_i, $i = 1, 2, ..., n$ are independent random variables with $E[X_i] = 0$, and $\sum_{i=1}^{n} \operatorname{Var}(X_i) = 1$. Let $W = \sum_{i=1}^{n} X_i$ and show that

$$\left| E|W| - \sqrt{\frac{2}{\pi}} \right| \leq 3 \sum_{i=1}^{n} E|X_i|^3.$$

3

Conditional Expectation and Martingales

3.1 Introduction

A generalization of a sequence of independent random variables occurs when we allow the variables in the sequence to be dependent on previous variables in the sequence. One example of this type of dependence is called a martingale, and its definition formalizes the concept of a fair gambling game. A number of results that hold for independent random variables also hold, under certain conditions, for martingales, and seemingly complex problems can be elegantly solved by reframing them in terms of a martingale.

In Section 3.2 we introduce the notion of conditional expectation, formally define a martingale in Section 3.3, introduce the concept of stopping times and prove the martingale stopping theorem in Section 3.4, give an approach for finding tail probabilities for martingale in Section 3.5, and introduce supermartingales and submartingales and prove the martingale convergence theorem in Section 3.6.

3.2 Conditional Expectation

Let X be such that $E[|X|] < \infty$. In a first course in probability, $E[X|Y]$, the conditional expectation of X given Y, is defined to be the function of Y that when $Y = y$ is equal to

$$E[X|Y = y] = \begin{cases} \sum_x x P(X = x | Y = y), \\ \qquad \text{if } X, Y \text{ are discrete} \\[2mm] \int x f_{X|Y}(x|y)dx, \\ \qquad \text{if } X, Y \text{ have joint density } f \end{cases}$$

where

$$f_{X|Y}(x|y) = \frac{f(x,y)}{\int f(x,y)dx} = \frac{f(x,y)}{f_Y(y)}. \tag{3.1}$$

The important result, often called the *tower property*,

$$E[X] = E[E[X|Y]],$$

is then proven. This result, which is often written as

$$E[X] = \begin{cases} \sum_y E[X|Y = y] P(Y = y) \\ \qquad \text{if } X, Y \text{ are discrete} \\[2mm] \int E[X|Y = y] f_Y(y)dy, \\ \qquad \text{if } X, Y \text{ are jointly continuous,} \end{cases}$$

is then gainfully employed in a variety of different calculations.

We now show how to give a more general definition of conditional expectation that reduces to the preceding cases when the random variables are discrete or continuous. To motivate our definition, suppose that whether or not A occurs is determined by the value of Y. (That is, suppose that $A \in \sigma(Y)$.) Then, using material from our first course in probability, we see that

$$\begin{aligned} E[X I_A] &= E[E[X I_A | Y]] \\ &= E[I_A E[X|Y]], \end{aligned}$$

where the final equality holds because, given Y, I_A is a constant random variable.

We are now ready for a general definition of conditional expectation.

Definition For random variables X, Y, let $E[X|Y]$, which is called the *conditional expectation of X given Y*, denote that function $h(Y)$ having the property that for any $A \in \sigma(Y)$

$$E[X I_A] = E[h(Y) I_A].$$

By the Radon–Nikodym theorem of measure theory, a function h that makes $h(Y)$ a (measurable) random variable and satisfies the preceding always exists, and as we show in the following, it is unique in the sense that any two such functions of Y must, with a probability of one, be equal. The function h is also referred to as a Radon–Nikodym derivative.

Proposition 3.1 *If h_1 and h_2 are functions such that*

$$E[h_1(Y)I_A] = E[h_2(Y)I_A]$$

for any $A \in \sigma(Y)$, then

$$P(h_1(Y) = h_2(Y)) = 1.$$

Proof Let $A_n = \{h_1(Y) - h_2(Y) > 1/n\}$. Then,

$$
\begin{aligned}
0 &= E[h_1(Y)I_{A_n}] - E[h_2(Y)I_{A_n}] \\
&= E[(h_1(Y) - h_2(Y))I_{A_n}] \\
&\geq \frac{1}{n}P(A_n),
\end{aligned}
$$

showing that $P(A_n) = 0$. Because the events A_n are increasing in n, this yields

$$0 = \lim_n P(A_n) = P(\lim_n A_n) = P(\cup_n A_n) = P(h_1(Y) > h_2(Y)).$$

Similarly, we can show that $0 = P(h_1(Y) < h_2(Y))$, which proves the result. ∎

We now show that the preceding general definition of conditional expectation reduces to the usual ones when X, Y are either jointly discrete or continuous.

Proposition 3.2 *If X and Y are both discrete, then*

$$E[X|Y = y] = \sum_x xP(X = x|Y = y),$$

whereas if they are jointly continuous with joint density f, then

$$E[X|Y = y] = \int x f_{X|Y}(x|y)dx,$$

where $f_{X|Y}(x|y) = \frac{f(x,y)}{\int f(x,y)dx}$.

Proof Suppose X and Y are both discrete, and let

$$h(y) = \sum_x xP(X = x|Y = y).$$

For $A \in \sigma(Y)$, define

$$B = \{y : I_A = 1 \text{ when } Y = y\}.$$

Because $B \in \sigma(Y)$, it follows that $I_A = I_B$. Thus,

$$XI_A = \begin{cases} x & \text{if } X = x, Y \in B \\ 0 & \text{otherwise} \end{cases}$$

and

$$h(Y)I_A = \begin{cases} \sum_x xP(X = x|Y = y) & \text{if } Y = y \in B. \\ 0 & \text{otherwise} \end{cases}$$

Thus,

$$\begin{aligned} E[XI_A] &= \sum_x xP(X = x, Y \in B) \\ &= \sum_x x \sum_{y \in B} P(X = x, Y = y) \\ &= \sum_{y \in B} \sum_x xP(X = x|Y = y)P(Y = y) \\ &= E[h(Y)I_A]. \end{aligned}$$

The result thus follows by uniqueness. The proof in the continuous case is similar and is left as an exercise. ∎

For any sigma field $\mathcal{F}$, we can define $E[X|\mathcal{F}]$ to be that random variable in $\mathcal{F}$ having the property that for all $A \in \mathcal{F}$

$$E[XI_A] = E[E[X|\mathcal{F}]I_A].$$

Intuitively, $E[X|\mathcal{F}]$ represents the conditional expectation of X given that we know all of the events in $\mathcal{F}$ that occur.

Remark 3.3 It follows from the preceding definition that $E[X|Y] = E[X|\sigma(Y)]$.

For any random variables $X, X_1, \ldots, X_n$, define $E[X|X_1, \ldots, X_n]$ by

$$E[X|X_1, \ldots, X_n] = E[X|\sigma(X_1, \ldots, X_n)].$$

In other words, $E[X|X_1, \ldots, X_n]$ is that function $h(X_1, \ldots, X_n)$ for which

$$E[XI_A] = E[h(X_1, \ldots, X_n)I_A], \qquad \text{for all } A \in \sigma(X_1, \ldots, X_n).$$

We now establish some important properties of conditional expectation.

Proposition 3.4

(a)   Tower property: For any sigma field $\mathcal{F}$

$$E[X] = E[E[X|\mathcal{F}]].$$

(b)   For any $A \in \mathcal{F}$,

$$E[XI_A|\mathcal{F}] = I_A E[X|\mathcal{F}].$$

(c) If X is independent of all $Y \in \mathcal{F}$, then

$$E[X|\mathcal{F}] = E[X].$$

(d) If $E\left[|X_i|\right] < \infty$, $i = 1, \ldots, n$, then

$$E\left[\sum_{i=1}^{n} X_i|\mathcal{F}\right] = \sum_{i=1}^{n} E[X_i|\mathcal{F}].$$

(e)   Jensen's inequality: If f is a convex function, then

$$E[f(X)|\mathcal{F}] \geq f(E[X|\mathcal{F}])$$

provided the expectations exist.

Proof   Recall that $E[X|\mathcal{F}]$ is the unique random variable in $\mathcal{F}$ such that

$$E[XI_A] = E[E[X|\mathcal{F}]I_A] \quad \text{if } A \in \mathcal{F}.$$

Letting $A = \Omega$, $I_A = I_\Omega = 1$, and Part (a) follows.

To prove Part (b), fix $A \in \mathcal{F}$ and let $X^* = XI_A$. Because $E[X^*|\mathcal{F}]$ is the unique function of Y such that $E[X^*I_{A'}] = E[E[X^*|\mathcal{F}]I_{A'}]$ for all $A' \in \mathcal{F}$, to show that $E[X^*|\mathcal{F}] = I_A E[X|\mathcal{F}]$, it suffices to show that for $A' \in \mathcal{F}$

$$E[X^*I_{A'}] = E[I_A E[X|\mathcal{F}]I_{A'}].$$

That is, it suffices to show that

$$E[XI_A I_{A'}] = E[I_A E[X|\mathcal{F}]I_{A'}]$$

or, equivalently, that

$$E[XI_{AA'}] = E[E[X|\mathcal{F}]I_{AA'}],$$

which because $AA' \in \mathcal{F}$, follows by the definition of conditional expectation.

Part (c) will follow if we can show that, for $A \in \mathcal{F}$

$$E[XI_A] = E[E[X]I_A],$$

which follows because

$$\begin{aligned} E[XI_A] &= E[X]E[I_A] \quad \text{by independence} \\ &= E[E[X]I_A]. \end{aligned}$$

We will leave the proofs of Parts (d) and (e) as exercises.

Remark 3.5 It can be shown that $E[X|\mathcal{F}]$ satisfies all the properties of ordinary expectations, except that all probabilities are now computed conditional on knowing which events in $\mathcal{F}$ have occurred. For instance, applying the tower property to $E[X|\mathcal{F}]$ yields

$$E[X|\mathcal{F}] = E[E[X|\mathcal{F} \cup \mathcal{G}]|\mathcal{F}].$$

Although the conditional expectation of X given the sigma field $\mathcal{F}$ is defined to be that function satisfying

$$E[XI_A] = E[E[X|\mathcal{F}]I_A] \quad \text{for all } A \in \mathcal{F},$$

it can be shown, using the dominated convergence theorem, that

$$E[XW] = E[E[X|\mathcal{F}]W] \quad \text{for all } W \in \mathcal{F}. \tag{3.2}$$

The following proposition is useful.

Proposition 3.6 *If* $W \in \mathcal{F}$, *then*

$$E[XW|\mathcal{F}] = WE[X|\mathcal{F}].$$

Before giving a proof, let us note that the result is intuitive. Because $W \in \mathcal{F}$, it follows that conditional on knowing which events of $\mathcal{F}$ occur (that is, conditional on $\mathcal{F}$), the random variable W becomes a constant, and the expected value of a constant times a random variable is just the constant times the expected value of the random variable. Next we formally prove the result.

Proof Let

$$\begin{aligned}
Y &= E[XW|\mathcal{F}] - WE[X|\mathcal{F}] \\
&= (X - E[X|\mathcal{F}])W - (XW - E[XW|\mathcal{F}]),
\end{aligned}$$

and note that $Y \in \mathcal{F}$. Now, for $A \in \mathcal{F}$,

$$E[YI_A] = E[(X - E[X|\mathcal{F}])WI_A] - E[(XW - E[XW|\mathcal{F}])I_A].$$

However, because $WI_A \in \mathcal{F}$, we use Equation 3.2 to get

$$E[(X - E[X|\mathcal{F}])WI_A] = E[XWI_A] - E[E[X|\mathcal{F}]WI_A] = 0,$$

and by the definition of conditional expectation, we have

$$E[(XW - E[XW|\mathcal{F}])I_A)] = E[XWI_A] - E[E[XW|\mathcal{F}]I_A] = 0.$$

Thus, we see that for $A \in \mathcal{F}$,

$$E[YI_A] = 0.$$

Setting first $A = \{Y > 0\}$, and then $A = \{Y < 0\}$ (which are both in $\mathcal{F}$ because $Y \in \mathcal{F}$) shows that

$$P(Y > 0) = P(Y < 0) = 0.$$

Hence, $Y = 0$, which proves the result. ∎

3.3 Martingales

We say that the sequence of sigma fields $\mathcal{F}_1, \mathcal{F}_2, \ldots$ is a *filtration* if $\mathcal{F}_1 \subseteq \mathcal{F}_2 \ldots$. We say a sequence of random variables $X_1, X_2, \ldots$ is adapted to $\mathcal{F}_n$ if $X_n \in \mathcal{F}_n$ for all n.

To obtain a feel for these definitions, it is useful to think of n as representing time, with information being gathered as time progresses. With this interpretation, the sigma field $\mathcal{F}_n$ represents all events that are determined by what occurs up to time n and thus contains $\mathcal{F}_{n-1}$. The sequence $X_n, n \geq 1$, is adapted to the filtration $\mathcal{F}_n, n \geq 1$, when the value of X_n is determined by what occurs by time n.

Definition 3.7 Z_n *is a martingale for filtration* $\mathcal{F}_n$ *if*

(a) $E[|Z_n|] < \infty$

(b) Z_n *is adapted to* $\mathcal{F}_n$

(c) $E[Z_{n+1}|\mathcal{F}_n] = Z_n$

A bet is said to be fair if its expected gain is equal to zero. A martingale can be thought of as being a generalized version of a fair game. For example, consider a gambling casino in which bets can be made concerning games played in sequence. Let $\mathcal{F}_n$ consist of all events with an outcome that is determined by the results of the first n games. Let Z_n denote the fortune of a specified gambler after the first n games. Then, the martingale condition states that regardless of the results of the first n games, the gambler's expected fortune after game $n+1$ is exactly what it was before that game. That is, no matter what has previously occurred, the gambler's expected winnings in any game is equal to zero.

It follows upon taking expectations of both sides of the final martingale condition that

$$E[Z_{n+1}] = E[Z_n],$$

implying that

$$E[Z_n] = E[Z_1].$$

We call $E[Z_1]$ the mean of the martingale.

Another useful martingale result is that

$$
\begin{aligned}
E[Z_{n+2}|\mathcal{F}_n] &= E[E[Z_{n+2}|\mathcal{F}_{n+1} \cup \mathcal{F}_n]|\mathcal{F}_n] \\
&= E[E[Z_{n+2}|\mathcal{F}_{n+1}]|\mathcal{F}_n] \\
&= E[Z_{n+1}|\mathcal{F}_n] \\
&= Z_n.
\end{aligned}
$$

Repeating this argument yields

$$E[Z_{n+k}|\mathcal{F}_n] = Z_n, \quad k \geq 1.$$

Definition 3.8 *We say that $Z_n, n \geq 1$, is a martingale (without specifying a filtration) if*

(a) $E[|Z_n|] < \infty$

(b) $E[Z_{n+1}|Z_1, \ldots, Z_n] = Z_n$

If $\{Z_n\}$ is a martingale for the filtration $\mathcal{F}_n$, then it is a martingale. This follows from

$$
\begin{aligned}
E[Z_{n+1}|Z_1, \ldots, Z_n] &= E[E[Z_{n+1}|Z_1, \ldots, Z_n, \mathcal{F}_n]|Z_1, \ldots, Z_n] \\
&= E[E[Z_{n+1}|\mathcal{F}_n]|Z_1, \ldots, Z_n] \\
&= E[Z_n|Z_1, \ldots, Z_n] \\
&= Z_n.
\end{aligned}
$$

where the second equality followed because $Z_i \in \mathcal{F}_n$, for all $i = 1, \ldots, n$.

We now give some examples of martingales.

Example 3.9 If $X_i, i \geq 1$, are independent zero mean random variables, then $Z_n = \sum_{i=1}^n X_i$, $n \geq 1$, is a martingale with respect to the filtration $\sigma(X_1, \ldots, X_n), n \geq 1$. This follows because

$$
\begin{aligned}
E[Z_{n+1}|X_1, \ldots, X_n] &= E[Z_n + X_{n+1}|X_1, \ldots, X_n] \\
&= E[Z_n|X_1, \ldots, X_n] + E[X_{n+1}|X_1, \ldots, X_n] \\
&= Z_n + E[X_{n+1}] \\
&\qquad \text{by the independence of the } X_i \\
&= Z_n. \quad \blacksquare
\end{aligned}
$$

Example 3.10 Let $X_i, i \geq 1$, be iid with mean zero and variance σ^2. Let $S_n = \sum_{i=1}^n X_i$, and define

$$
Z_n = S_n^2 - n\sigma^2, \quad n \geq 1.
$$

Then $Z_n, n \geq 1$ is a martingale for the filtration $\sigma(X_1, \ldots, X_n)$. To verify this claim, note that

$$
\begin{aligned}
E[S_{n+1}^2|X_1, \ldots, X_n] &= E[(S_n + X_{n+1})^2|X_1, \ldots, X_n] \\
&= E[S_n^2|X_1, \ldots, X_n] + E[2S_n X_{n+1}|X_1, \ldots, X_n] \\
&\quad + E[X_{n+1}^2|X_1, \ldots, X_n] \\
&= S_n^2 + 2S_n E[X_{n+1}|X_1, \ldots, X_n] + E[X_{n+1}^2] \\
&= S_n^2 + 2S_n E[X_{n+1}] + \sigma^2 \\
&= S_n^2 + \sigma^2.
\end{aligned}
$$

Subtracting $(n+1)\sigma^2$ from both sides, yields

$$
E[Z_{n+1}|X_1, \ldots, X_n] = Z_n. \quad \blacksquare
$$

Our next example introduces an important type of martingale, known as a Doob martingale.

Example 3.11 Let Y be an arbitrary random variable with $E[|Y|] < \infty$, let $\mathcal{F}_n, n \geq 1$, be a filtration, and define

$$Z_n = E[Y|\mathcal{F}_n].$$

We claim that $Z_n, n \geq 1$, is a martingale with respect to the filtration $\mathcal{F}_n, n \geq 1$. To verify this, we must first show that $E[|Z_n|] < \infty$, which is accomplished as follows:

$$
\begin{aligned}
E[|Z_n|] &= E[|E[Y|\mathcal{F}_n]|] \\
&\leq E[E[|Y||\mathcal{F}_n]] \\
&= E[|Y|] \\
&< \infty,
\end{aligned}
$$

where the first inequality uses that the function $f(x) = |x|$ is convex, and thus from Jensen's inequality

$$E[|Y||\mathcal{F}_n] \geq |E[Y|\mathcal{F}_n]|,$$

whereas the final equality used the tower property. The verification is now completed as follows:

$$
\begin{aligned}
E[Z_{n+1}|\mathcal{F}_n] &= E[E[Y|\mathcal{F}_{n+1}]|\mathcal{F}_n] \\
&= E[Y|\mathcal{F}_n] \qquad \text{by the tower property} \\
&= Z_n.
\end{aligned}
$$

The martingale $Z_n, n \geq 1$, is a called a *Doob martingale*. ∎

Example 3.12 Our final example generalizes the result that the successive partial sums of independent zero mean random variables constitute a martingale. For any random variables $X_i, i \geq 1$, the random variables $X_i - E[X_i|X_1, \ldots, X_{i-1}], i \geq 1$, have mean zero. Even though they need not be independent, their partial sums constitute a martingale. That is, we claim that

$$Z_n = \sum_{i=1}^{n}(X_i - E[X_i|X_1, \ldots, X_{i-1}]), \; n \geq 1,$$

is, provided that $E[|Z_n|] < \infty$, a martingale with respect to the filtration $\sigma(X_1, \ldots, X_n), n \geq 1$. To verify this claim, note that

$$Z_{n+1} = Z_n + X_{n+1} - E[X_{n+1}|X_1, \ldots, X_n].$$

Thus,

$$E[Z_{n+1}|X_1, \ldots, X_n] = Z_n + E[X_{n+1}|X_1, \ldots, X_n]$$
$$- E[X_{n+1}|X_1, \ldots, X_n]$$
$$= Z_n. \quad \blacksquare$$

3.4   Martingale Stopping Theorem

The positive integer valued, possibly infinite, random variable N is said to be a *random time* for the filtration $\mathcal{F}_n$ if $\{N = n\} \in \mathcal{F}_n$ or, equivalently, if $\{N \geq n\} \in \mathcal{F}_{n-1}$, for all n. If $P(N < \infty) = 1$, then the random time is said to be a *stopping time* for the filtration.

Thinking once again in terms of information being amassed over time, with $\mathcal{F}_n$ being the cumulative information for all events that have occurred by time n, the random variable N will be a stopping time for this filtration if the decision whether to stop at time n (so that $N = n$) depends only on what has occurred by time n. (That is, the decision to stop at time n is not allowed to depend on the results of future events.) It should be noted, however, that the decision to stop at time n need not be independent of future events, only that it must be conditionally independent of future events given all information up to the present.

Lemma 3.13 *Let* $Z_n, n \geq 1$, *be a martingale for the filtration* $\mathcal{F}_n$. *If* N *is a random time for this filtration, then the process* $\bar{Z}_n = Z_{\min(N,n)}$, $n \geq 1$, *called the stopped process, is also a martingale for the filtration* $\mathcal{F}_n$.

Proof Start with the identity

$$\bar{Z}_n = \bar{Z}_{n-1} + I_{\{N \geq n\}}(Z_n - Z_{n-1}).$$

To verify the preceding, consider two cases:

1. $N \geq n$: Here, $\bar{Z}_n = Z_n$, $\bar{Z}_{n-1} = Z_{n-1}$, $I_{\{N \geq n\}} = 1$, and the preceding is true.

2. $N < n$: Here, $\bar{Z}_n = \bar{Z}_{n-1} = Z_N$, $I_{\{N \geq n\}} = 0$, and the preceding is true.

Hence,

$$E[\bar{Z}_n|\mathcal{F}_{n-1}] = E[\bar{Z}_{n-1}|\mathcal{F}_{n-1}] + E[I_{\{N \geq n\}}(Z_n - Z_{n-1})|\mathcal{F}_{n-1}]$$
$$= \bar{Z}_{n-1} + I_{\{N \geq n\}}E[Z_n - Z_{n-1}|\mathcal{F}_{n-1}]$$
$$= \bar{Z}_{n-1}. \quad \blacksquare$$

Theorem 3.14 *Martingale stopping theorem.*   *Let* Z_n, $n \geq 1$, *be a martingale for the filtration* $\mathcal{F}_n$, *and suppose that* N *is a stopping time for this filtration. Then*

$$E[Z_N] = E[Z_1]$$

if any of the following three sufficient conditions hold.
(a) $\bar{Z}_n$ are uniformly bounded;
(b) N is bounded; or
(c) $E[N] < \infty$, and there exists $M < \infty$ such that

$$E[|Z_{n+1} - Z_n||\mathcal{F}_n] < M.$$

Proof Because the stopped process is also a martingale,

$$E[\bar{Z}_n] = E[\bar{Z}_1] = E[Z_1].$$

Because $P(N < \infty) = 1$, it follows that $\bar{Z}_n = Z_N$ for sufficiently large N, implying that

$$\lim_{n \to \infty} \bar{Z}_n = Z_N.$$

Part (a) follows from the bounded convergence theorem, and Part (b) with the bound $N \le n$ follows from the dominated convergence theorem using the bound $|\bar{Z}_j| \le \sum_{i=1}^{n} |\bar{Z}_i|$.

To prove Part (c) note that with $\bar{Z}_0 = 0$

$$
\begin{aligned}
|\bar{Z}_n| &= \left| \sum_{i=1}^{n} (\bar{Z}_i - \bar{Z}_{i-1}) \right| \\
&\le \sum_{i=1}^{\infty} |\bar{Z}_i - \bar{Z}_{i-1}| \\
&= \sum_{i=1}^{\infty} I_{\{N \ge i\}} |Z_i - Z_{i-1}|,
\end{aligned}
$$

and the result now follows from the dominated convergence theorem because

$$
\begin{aligned}
E\left[\sum_{i=1}^{\infty} I_{\{N \ge i\}} |Z_i - Z_{i-1}| \right] &= \sum_{i=1}^{\infty} E[I_{\{N \ge i\}} |Z_i - Z_{i-1}|] \\
&= \sum_{i=1}^{\infty} E[E[I_{\{N \ge i\}} |Z_i - Z_{i-1}||\mathcal{F}_{i-1}]] \\
&= \sum_{i=1}^{\infty} E[I_{\{N \ge i\}} E[|Z_i - Z_{i-1}||\mathcal{F}_{i-1}]] \\
&\le M \sum_{i=1}^{\infty} P(N \ge i) \\
&= M E[N] \\
&< \infty.
\end{aligned}
$$

∎

A corollary of the martingale stopping theorem is Wald's equation.

Corollary 3.15 *Wald's equation. If $X_1, X_2, \ldots$ are iid with finite mean $\mu = E[X_i]$, and if N is a stopping time for the filtration $\mathcal{F}_n = \sigma(X_1, \ldots, X_n), n \geq 1$, such that $E[N] < \infty$, then*

$$E\left[\sum_{i=1}^{N} X_i\right] = \mu E[N].$$

Proof $Z_n = \sum_{i=1}^{n}(X_i - \mu), n \geq 1$, being the successive partial sums of independent zero mean random variables, is a martingale with mean zero. Hence, assuming the martingale stopping theorem can be applied we have that

$$0 = E[Z_N]$$
$$= E\left[\sum_{i=1}^{N}(X_i - \mu)\right]$$
$$= E\left[\sum_{i=1}^{N} X_i - N\mu)\right]$$
$$= E\left[\sum_{i=1}^{N} X_i\right] - E[N\mu].$$

To complete the proof, we verify the sufficient condition from Part (c) of the martingale stopping theorem.

$$E[|Z_{n+1} - Z_n| | \mathcal{F}_n] = E[|X_{n+1} - \mu| | \mathcal{F}_n]$$
$$= E[|X_{n+1} - \mu|] \quad \text{by independence.}$$
$$\leq E[|X_i|] + |\mu|$$

Thus, the condition from Part (c) is verified and the result proved. ∎

Example 3.16 Suppose iid discrete random variables $X_1, X_2, \ldots$ are observed in sequence. With $P(X_i = j) = p_j$, what is the expected number of random variables that must be observed until the subsequence $0, 1, 2, 0, 1$ occurs? What is the variance?

Solution   Consider a fair gambling casino in which the expected casino win for every bet is zero. Note that if a gambler bets their entire fortune of a that the next outcome is j; then their fortune after the bet will either be 0 with probability $1 - p_j$, or a/p_j with probability p_j. Now, imagine a sequence of gamblers betting at this casino. Each gambler starts with an initial fortune of one and stops playing if their fortune ever becomes zero. Gambler i bets one that $X_i = 0$. If they win, they bet their entire fortune (of $1/p_0$) that $X_{i+1} = 1$; if they win that bet, they bet their entire fortune that $X_{i+2} = 2$; if they win that bet, they bet their entire fortune that $X_{i+3} = 0$; if they win that bet, they bet their entire fortune that $X_{i+4} = 1$; and if they win that bet, they quit with a final fortune of $(p_0^2 p_1^2 p_2)^{-1}$.

Let Z_n denote the casino's winnings after the data value X_n is observed; because it is a fair casino, $Z_n, n \geq 1$, is a martingale with mean zero with respect to the filtration $\sigma(X_1, \ldots, X_n), n \geq 1$. Let N denote the number of random variables that need be observed until the pattern $0, 1, 2, 0, 1$ appears – so $(X_{N-4}, \ldots, X_N) = (0, 1, 2, 0, 1)$. Because it is easy to verify that N is a stopping time for the filtration, and that the condition in Part (c) of the martingale stopping theorem is satisfied when $M = 4/(p_0^2 p_1^2 p_2)$, it follows that $E[Z_N] = 0$. However, after X_N has been observed, each of the gamblers $1, \ldots, N - 5$ would have lost one: Gambler $N - 4$ would have won $(p_0^2 p_1^2 p_2)^{-1} - 1$; gamblers $N - 3$ and $N - 2$ would each have lost one; gambler $N - 1$ would have won $(p_0 p_1)^{-1} - 1$; and gambler N would have lost one. Therefore,

$$Z_N = N - (p_0^2 p_1^2 p_2)^{-1} - (p_0 p_1)^{-1}.$$

Using $E[Z_N] = 0$ yields the result

$$E[N] = (p_0^2 p_1^2 p_2)^{-1} + (p_0 p_1)^{-1}.$$

In the same manner, we can compute the expected time until any specified pattern occurs in iid generated random data. For instance, when making independent flips of a coin that comes up heads with probability p, the mean number of flips until the pattern $HHTTHH$ appears is $p^{-4} q^{-2} + p^{-2} + p^{-1}$, where $q = 1 - p$.

To determine $\mathrm{Var}(N)$, suppose now that gambler i starts with an initial fortune i and bets that amount that $X_i = 0$. If they win, they bet their entire fortune that $X_{i+1} = 1$; if they win that bet, they bet their entire fortune that $X_{i+2} = 2$; if they win that bet, they bet their entire fortune that $X_{i+3} = 0$; if they win that bet, they bet their entire fortune that $X_{i+4} = 1$; and if they win that bet, they quit with a final fortune of $i/(p_0^2 p_1^2 p_2)$. The casino's winnings at time N is thus

$$\begin{aligned} Z_N &= 1 + 2 + \cdots + N - \frac{N-4}{p_0^2 p_1^2 p_2} - \frac{N-1}{p_0 p_1} \\ &= \frac{N(N+1)}{2} - \frac{N-4}{p_0^2 p_1^2 p_2} - \frac{N-1}{p_0 p_1}. \end{aligned}$$

Assuming that the martingale stopping theorem holds (although none of the three sufficient conditions hold, the stopping theorem can still be shown to be valid for this martingale), we obtain upon taking expectations that

$$E[N^2] + E[N] = 2\frac{E[N] - 4}{p_0^2 p_1^2 p_2} + 2\frac{E[N] - 1}{p_0 p_1}.$$

Using the previously obtained value of $E[N]$, the preceding can now be solved for $E[N^2]$ to obtain $\text{Var}(N) = E[N^2] - (E[N])^2$. ∎

Example 3.17 The cards from a shuffled deck of 26 red and 26 black cards are to be turned over one at a time. At any time, a player can request the next card and is a winner if the next card is red and is a loser otherwise. A player who has not yet requested another card when only a single card remains is a winner if the final card is red and is a loser otherwise. What is a good strategy for the player?

Solution Every strategy has probability $1/2$ of resulting in a win. To see this, let R_n denote the number of red cards remaining in the deck after n cards have been shown. Then

$$E[R_{n+1}|R_1, \ldots, R_n] = R_n - \frac{R_n}{52 - n} = \frac{51 - n}{52 - n}R_n.$$

Hence, $\frac{R_n}{52-n}$, $n \geq 0$ is a martingale. Because $R_0/52 = 1/2$, this martingale has mean $1/2$. Now, consider any strategy, and let N denote the number of cards that are turned over before the next card is requested. Because N is a bounded stopping time, it follows from the martingale stopping theorem that

$$E\left[\frac{R_N}{52 - N}\right] = 1/2.$$

Hence, with $I = I_{\{\text{win}\}}$

$$E[I] = E[E[I|R_N]] = E\left[\frac{R_N}{52 - N}\right] = 1/2. \quad \blacksquare$$

Our next example involves the matching problem.

Example 3.18 Each member of a group of n individuals throws their hat in a pile. The hats are then mixed together, and each person randomly selects a hat in such a manner that each of the $n!$ possible selections of the n individuals are equally likely. Any individual who selects their own hat departs, and that is the end of round one. If any individuals remain, then each throws the hat they have in a pile, and each one then randomly chooses a hat. Those selecting their own hats leave, and that ends round two. Find $E[N]$, where N is the number of rounds until everyone has departed.

Solution   Let $X_i, i \geq 1$, denote the number of matches on round i for $i = 1, \ldots, N$, and let it equal one for $i > N$. To solve this example, we will use the zero mean martingale $Z_k, k \geq 1$, defined by

$$
\begin{aligned}
Z_k &= \sum_{i=1}^{k} (X_i - E[X_i | X_1, \ldots, X_{i-1}]) \\
&= \sum_{i=1}^{k} (X_i - 1),
\end{aligned}
$$

where the final equality follows because, for any number of remaining individuals, the expected number of matches in a round is one (which is seen by writing this as the sum of indicator variables for the events that each remaining person has a match). Because

$$
N = \min \left\{ k : \sum_{i=1}^{k} X_i = n \right\}
$$

is a stopping time for this martingale, we obtain from the martingale stopping theorem that

$$
0 = E[Z_N] = E \left[\sum_{i=1}^{N} X_i - N \right] = n - E[N],
$$

so $E[N] = n$.   ∎

Example 3.19 If $X_1, X_2, \ldots$, is a sequence of iid random variables, $P(X_i = 0) \neq 1$, then the process

$$
S_n = \sum_{i=1}^{n} X_i, \quad i \geq 1
$$

is said to be a *random walk*. For given positive constants a, b, let p denote the probability that the the random walk becomes as large as a before it becomes as small as $-b$.

We now show how to use martingale theory to approximate p. In the case where we have a bound $|X_i| < c$, it can be shown there will be a value $\theta \neq 0$ such that

$$
E[e^{\theta X}] = 1.
$$

Then, because

$$
Z_n = e^{\theta S_n} = \prod_{i=1}^{n} e^{\theta X_i}
$$

is the product of independent random variables with mean one, it follows that $Z_n, n \geq 1$, is a martingale having mean one. Let

$$N = \min(n : S_n \geq a \quad \text{or} \quad S_n \leq -b).$$

The condition in Part (c) of the martingale stopping theorem can be shown to hold, implying that

$$E[e^{\theta S_N}] = 1.$$

Thus,

$$1 = E[e^{\theta S_N}|S_N \geq a]p + E[e^{\theta S_N}|S_N \leq -b](1-p).$$

Now, if $\theta > 0$, then

$$e^{\theta a} \leq E[e^{\theta S_N}|S_N \geq a] \leq e^{\theta(a+c)}$$

and

$$e^{-\theta(b+c)} \leq E[e^{\theta S_N}|S_N \leq -b] \leq e^{-\theta b},$$

yielding the bounds

$$\frac{1 - e^{-\theta b}}{e^{\theta(a+c)} - e^{-\theta b}} \leq p \leq \frac{1 - e^{-\theta(b+c)}}{e^{\theta a} - e^{-\theta(b+c)}}$$

and motivating the approximation

$$p \approx \frac{1 - e^{-\theta b}}{e^{\theta a} - e^{-\theta b}}.$$

We leave it as an exercise to obtain bounds on p when $\theta < 0$. ∎

Our next example involves a Doob backward martingale. Before defining this martingale, we need the following definition.

Definition 3.20 *The random variables $X_1, \ldots, X_n$ are said to be exchangeable if $X_{i_1}, \ldots, X_{i_n}$ has the same probability distribution for every permutation $i_1, \ldots, i_n$ of $1, \ldots, n$.*

Suppose that $X_1, \ldots, X_n$ are exchangeable. Assume $E[|X_1|] < \infty$, let

$$S_j = \sum_{i=1}^{j} X_i, \quad j = 1, \ldots, n$$

and consider the Doob martingale $Z_1, \ldots, Z_n$, given by

$$Z_1 = E[X_1|S_n]$$
$$Z_j = E[X_1|S_n, S_{n-1}, \ldots, S_{n+1-j}], \quad j = 1, \ldots, n.$$

However,

$$\begin{aligned} S_{n+1-j} &= E[S_{n+1-j}|S_{n+1-j}, X_{n+2-j}, \ldots, X_n] \\ &= \sum_{i=1}^{n+1-j} E[X_i|S_{n+1-j}, X_{n+2-j}, \ldots, X_n] \\ &= (n+1-j)E[X_1|S_{n+1-j}, X_{n+2-j}, \ldots, X_n] \\ &\quad \text{(by exchangeability)} \\ &= (n+1-j)Z_j, \end{aligned}$$

where the final equality follows because knowing $S_n, S_{n-1}, \ldots, S_{n+1-j}$ is equivalent to knowing $S_{n+1-j}, X_{n+2-j}, \ldots, X_n$.

The martingale

$$Z_j = \frac{S_{n+1-j}}{n+1-j}, \ j = 1, \ldots, n$$

is called the *Doob backward martingale*. We now apply it to solve the ballot problem.

Example 3.21 In an election between candidates A and B, candidate A receives n votes and candidate B receives m votes, where $n > m$. Assuming that all orderings of the $n+m$ votes are equally likely, what is the probability that A is always ahead in the count of the votes?

Solution Let X_i equal one if the ith voted counted is for A, and let it equal -1 if that vote is for B. Because all orderings of the $n+m$ votes are assumed to be equally likely, it follows that the random variables $X_1, \ldots, X_{n+m}$ are exchangeable, and $Z_1, \ldots, Z_{n+m}$ is a Doob backward martingale when

$$Z_j = \frac{S_{n+m+1-j}}{n+m+1-j},$$

where $S_k = \sum_{i=1}^k X_i$. Because $Z_1 = S_{n+m}/(n+m) = (n-m)/(n+m)$, the mean of this martingale is $(n-m)/(n+m)$. Because $n > m$, candidate A will always be ahead in the count of the vote unless there is a tie at some point, which will occur if one of the S_j (or equivalently, one of the Z_j) is equal to zero. Consequently, define the bounded stopping time N by

$$N = \min\{j : Z_j = 0 \text{ or } j = n+m\}.$$

Because $Z_{n+m} = X_1$, it follows that Z_N will equal zero if the candidates are ever tied and will equal X_1 if A is always ahead. However, if A is always ahead, then A must receive the first vote; therefore,

$$Z_N = \begin{cases} 1 & \text{if A is always ahead} \\ 0 & \text{otherwise.} \end{cases}$$

By the martingale stopping theorem, $E[Z_N] = (n-m)/(n+m)$, yielding the result

$$P(\text{A is always ahead}) = \frac{n-m}{n+m}. \qquad \blacksquare$$

Example 3.22 *Ante one game.* There are three players, with player i initially having a fortune $x_i \geq 0$, $i = 1, 2, 3$. Say that a player is alive if their current fortune is positive. At the beginning of a round, all of the alive players put one into a pot, which is then equally likely to be won by any of these players. Let G_i be the number of games that involve i players. We now show how to compute $E[G_2]$ and $E[G_3]$.

Let $X_i(n)$ be player i's fortune after game n, and let $W_i(n)$ be player i's winnings in game n. Also let $I_i(n) = I\{X_i(n) > 0\}$ be the indicator of the event that i is still alive after game n, and $N(n) = \sum_{i=1}^{k} I_i(n)$ be the number of players that are still alive after game n. Then

$$W_i(n+1) = I_i(n)(N(n)J_i(n+1) - 1),$$

where $J_i(n+1)$, the indicator of the event that i wins game $n+1$, is such that

$$P(J_i(n+1) = 1|F_n) = I_i(n)\frac{1}{N(n)} = 1 - P(J_i(n+1) = 0|F_n),$$

where F_n is the sigma field generated by everything that happens up to and including game n. Hence,

$$\begin{aligned}
E[W_i(n+1)|F_n] &= I_i(n)(1-1) = 0 \\
E[W_i^2(n+1)|F_n] &= I_i(n)\, E[N^2(n)J_i(n+1) - 2N(n)J_i(n+1) + 1|F_n] \\
&= I_i(n)(N(n) - 1)
\end{aligned}$$

$$\begin{aligned}
E[W_i^3&(n+1)|F_n] \\
&= I_i(n)\, E[N^3(n)J_i(n+1) - 3N^2(n)J_i(n+1) + 3N(n)J_i(n+1) - 1|F_n] \\
&= I_i(n)(N^2(n) - 3N(n) + 3 - 1) \\
&= I_i(n)(N(n) - 1)(N(n) - 2).
\end{aligned}$$

Using

$$X_i(n+1) = X_i(n) + W_i(n+1),$$

the preceding gives

$$\begin{aligned}
E[X_i(n+1)|F_n] &= X_i(n) \\
E[X_i^2(n+1)|F_n] &= X_i^2(n) + 2X_i(n)E[W_i(n+1)|F_n] + E[W_i^2(n+1)|F_n] \\
&= X_i^2(n) + I_i(n)(N(n) - 1)
\end{aligned}$$

$$E[X_i^3(n+1)|F_n]$$
$$= X_i^3(n) + 3X_i^2(n)E[W_i(n+1)|F_n] + 3X_i(n)E[W_i^2(n+1)|F_n]$$
$$+ E[W_i^3(n+1)|F_n]$$
$$= X_i^3(n) + 3X_i(n)I_i(n)(N(n)-1) + I_i(n)(N(n)-1)(N(n)-2).$$

With $s = x_1 + x_2 + x_3$, it follows from the preceding that

$$E\left[\sum_{i=1}^3 X_i(n+1)|F_n\right] = \sum_{i=1}^3 X_i(n) = s$$

$$E\left[\sum_{i=1}^3 X_i^2(n+1)|F_n\right] = \sum_{i=1}^3 X_i^2(n) + N(n)(N(n)-1)$$

$$E\left[\sum_{i=1}^3 X_i^3(n+1)|F_n\right]$$

$$= \sum_{i=1}^3 X_i^3(n) + 3(N(n)-1)\sum_{i=1}^3 X_i(n)I_i(n) + N(n)(N(n)-1)(N(n)-2)$$

$$= \sum_{i=1}^k X_i^3(n) + 3(N(n)-1)s + N(n)(N(n)-1)(N(n)-2),$$

where the preceding used that $\sum_{i=1}^3 X_i(n)I_i(n) = \sum_{i=1}^3 X_i(n) = s$.

Now, if

$$E[V_{n+1}|F_n] = V_n + Y_n, \ n \geq 0,$$

then $\sum_{j=1}^n (V_j - E[V_j|F_{j-1}]) = V_n - V_0 - \sum_{j=0}^{n-1} Y_j$ is a zero mean martingale, which implies that

$$Z_n = V_n - \sum_{j=0}^{n-1} Y_j, \ n \geq 1$$

is a martingale with mean V_0. Consequently, both

$$Z_1(n) = \sum_{i=1}^3 X_i^2(n) - \sum_{j=0}^{n-1} N(j)(N(j)-1), \ n \geq 0$$

and

$$Z_2(n) = \sum_{i=1}^3 X_i^3(n) - 3s\sum_{j=0}^{n-1}(N(j)-1) - \sum_{j=0}^{n-1} N(j)(N(j)-1)(N(j)-2), \ n \geq 0$$

are martingales, with $Z_1(n)$ having mean $\sum_{i=1}^3 x_i^2$ and $Z_2(n)$ having mean $\sum_{i=1}^3 x_i^3$. Letting T be the number of games until one of the player's fortune is s, and using that $\sum_{i=1}^{3} X_i^j(T) = s^j$, $j = 2, 3$, the martingale stopping theorem gives that

$$\sum_{i=1}^{3} x_i^2 = E[Z_1(T)] = s^2 - E\left[\sum_{j=0}^{T-1} N(j)(N(j) - 1)\right]$$

$$\sum_{i=1}^{k} x_i^3 = E[Z_2(T)] = s^3 - 3s\,E\left[\sum_{j=0}^{T-1} (N(j) - 1)\right]$$

$$-E\left[\sum_{j=0}^{T-1} N(j)(N(j) - 1)(N(j) - 2)\right].$$

Now, let G_i be the number of games involving i players, $i = 2, 3$. Because $N(j)(N(j)-1)$ will equal $i(i-1)$ if game j involves i players, it follows that $\sum_{j=0}^{T-1} N(j)(N(j) - 1) = \sum_{i=2}^{3} i(i-1)G_i$, and similarly that $\sum_{j=0}^{T-1} (N(j) - 1) = \sum_{i=2}^{3} (i - 1)G_i$ and $\sum_{j=0}^{T-1} N(j)(N(j) - 1)(N(j) - 2) = \sum_{i=2}^{3} i(i - 1)(i - 2)G_i = 6G_3$. Hence, from the preceding we see that

$$\sum_{i=1}^{3} x_i^2 = s^2 - 2E[G_2] - 6E[G_3]$$

$$\sum_{i=1}^{3} x_i^3 = s^3 - 3sE[G_2] - (6s + 6)E[G_3].$$

Solving these equations gives

$$E[G_2] = \frac{\sum_{i=1}^{3} x_i(x_i - 1)(s - x_i)}{s - 2}$$

and

$$E[G_3] = \frac{x_1 x_2 x_3}{s - 2}.$$

Hence,

$$E[T] = E[G_2 + G_3] = \frac{x_1 x_2 x_3 + \sum_{i=1}^{3} x_i(x_i - 1)(s - x_i)}{s - 2}.$$

Remark 3.23 The following are two remarks for the previous example:

1. The martingale stopping theorem applies in the preceding because Markov chain theory shows that $E[T] < \infty$.

2. If we let P_i be the probability that player i eventually has fortune s, then using that $X_i(n), n \geq 0$ is a martingale with mean x_i, it follows from the martingale stopping theorem that $x_i = E[X_i(T)] = sP_i$, giving $P_i = \frac{x_i}{s}$.

3.5 Hoeffding–Azuma Inequality

Let $Z_n, n \geq 0$, be a martingale with respect to the filtration $\mathcal{F}_n$. If the differences $Z_n - Z_{n-1}$ can be shown to lie in a bounded random interval of the form $[-B_n, -B_n + d_n]$, where $B_n \in \mathcal{F}_{n-1}$ and d_n is constant, then the Hoeffding–Azuma inequality often yields useful bounds on the tail probabilities of Z_n. Before presenting the inequality, we will need a couple of lemmas.

Lemma 3.24 *If $E[X] = 0$, and $P(-\alpha \leq X \leq \beta) = 1$, then for any convex function f*

$$E[f(X)] \leq \frac{\beta}{\alpha + \beta} f(-\alpha) + \frac{\alpha}{\alpha + \beta} f(\beta).$$

Proof Because f is convex it follows that, in the region $-\alpha \leq x \leq \beta$, it is never above the line segment connecting the points $(-\alpha, f(-\alpha))$ and $(\beta, f(\beta))$. Because the formula for this line segment is

$$y = \frac{\beta}{\alpha + \beta} f(-\alpha) + \frac{\alpha}{\alpha + \beta} f(\beta) + \frac{1}{\alpha + \beta} [f(\beta) - f(-\alpha)] x,$$

we obtain from the condition $P(-\alpha \leq X \leq \beta) = 1$ that

$$f(X) \leq \frac{\beta}{\alpha + \beta} f(-\alpha) + \frac{\alpha}{\alpha + \beta} f(\beta) + \frac{1}{\alpha + \beta} [f(\beta) - f(-\alpha)] X.$$

Taking expectations, and using $E[X] = 0$, yields the result. ∎

Lemma 3.25 *For $0 \leq p \leq 1$*

$$p e^{t(1-p)} + (1 - p) e^{-tp} \leq e^{t^2/8}.$$

Proof Letting $p = (1 + \alpha)/2$ and $t = 2\beta$, we must show that for $-1 \leq \alpha \leq 1$

$$(1 + \alpha) e^{\beta(1-\alpha)} + (1 - \alpha) e^{-\beta(1+\alpha)} \leq 2 e^{\beta^2/2}$$

or, equivalently,

$$e^{\beta} + e^{-\beta} + \alpha(e^{\beta} - e^{-\beta}) \leq 2 e^{\alpha\beta + \beta^2/2}.$$

The preceding inequality is true when $\alpha = -1$ or $+1$ and when $|\beta|$ is large (say, when $|\beta| \geq 100$). Thus, if the Lemma were false, then the function

$$f(\alpha, \beta) = e^{\beta} + e^{-\beta} + \alpha(e^{\beta} - e^{-\beta}) - 2 e^{\alpha\beta + \beta^2/2}$$

twould assume a strictly positive maximum in the interior of the region $R = \{(\alpha, \beta) : |\alpha| \leq 1, |\beta| \leq 10\}$. Setting the partial derivatives of f equal to 0, we obtain

$$e^\beta - e^{-\beta} + \alpha(e^\beta + e^{-\beta}) = 2\alpha\beta e^{\alpha\beta + \beta^2/2} \qquad (3.3)$$

$$e^\beta - e^{-\beta} = 2\beta e^{\alpha\beta + \beta^2/2}. \qquad (3.4)$$

We will now prove the lemma by showing that any solution of Equations 3.3 and 3.4 must have $\beta = 0$. However, because $f(\alpha, 0) = 0$, this would contradict the hypothesis that f assumes a strictly positive maximum in R, thus establishing the lemma.

So, assume that there is a solution of Equations 3.3 and 3.4 in which $\beta \neq 0$. Now note that there is no solution of these equations for which $\alpha = 0$ and $\beta \neq 0$. For if there were such a solution, then Equation 3.4 would say that

$$e^\beta - e^{-\beta} = 2\beta e^{\beta^2/2}, \qquad (3.5)$$

But expanding in a power series about zero shows that Equation 3.5 is equivalent to

$$2 \sum_{i=0}^\infty \frac{\beta^{2i+1}}{(2i + 1)!} = 2 \sum_{i=0}^\infty \frac{\beta^{2i+1}}{i!2^i},$$

which (because $(2i + 1)! > i!2^i$ when $i > 0$) is clearly impossible when $\beta \neq 0$. Thus, any solution of Equations 3.3 and 3.4 in which $\beta \neq 0$ will also have $\alpha \neq 0$. Assuming such a solution gives, upon dividing Equation 3.3 by Equation 3.4,

$$1 + \alpha \frac{e^\beta + e^{-\beta}}{e^\beta - e^{-\beta}} = 1 + \frac{\alpha}{\beta}.$$

Because $\alpha \neq 0$, the preceding is equivalent to

$$\beta(e^\beta + e^{-\beta}) = e^\beta - e^{-\beta}.$$

or, expanding in a Taylor series,

$$\sum_{i=0}^\infty \frac{\beta^{2i+1}}{(2i)!} = \sum_{i=0}^\infty \frac{\beta^{2i+1}}{(2i + 1)!},$$

which is clearly not possible when $\beta \neq 0$. Thus, there is no solution of Equations 3.3 and 3.4 in which $\beta \neq 0$, thus proving the result.   ∎

We are now ready for the Hoeffding–Azuma inequality.

Theorem 3.26   *Hoeffding–Azuma inequality.*   *Let Z_n, $n \geq 1$, be a martingale with mean μ with respect to the filtration $\mathcal{F}_n$. Let $Z_0 = \mu$ and suppose there exist nonnegative random variables $B_n, n > 0$, where $B_n \in \mathcal{F}_{n-1}$, and positive constants $d_n, n > 0$, such that*

$$-B_n \leq Z_n - Z_{n-1} \leq -B_n + d_n.$$

Then, for $n > 0, a > 0$,

$$(i) \quad P(Z_n - \mu \geq a) \ \leq \ e^{-2a^2/\sum_{i=1}^{n} d_i^2}$$

$$(ii) \quad P(Z_n - \mu \leq -a) \ \leq \ e^{-2a^2/\sum_{i=1}^{n} d_i^2}. \tag{3.6}$$

Proof   Suppose that $\mu = 0$. For any $c > 0$,

$$P(Z_n \geq a) \ = \ P(e^{cZ_n} \geq e^{ca})$$
$$\leq \ e^{-ca} E[e^{cZ_n}],$$

where we use Markov's inequality in the second equality. Let $W_n = e^{cZ_n}$. Note that $W_0 = 1$ and that for $n > 0$

$$E[W_n|\mathcal{F}_{n-1}] \ = \ E[e^{cZ_{n-1}} e^{c(Z_n-Z_{n-1})}|\mathcal{F}_{n-1}]$$
$$= \ e^{cZ_{n-1}} E[e^{c(Z_n-Z_{n-1})}|\mathcal{F}_{n-1}]$$
$$= \ W_{n-1} E[e^{c(Z_n-Z_{n-1})}|\mathcal{F}_{n-1}], \tag{3.7}$$

where the second equality used that $Z_{n-1} \in \mathcal{F}_{n-1}$. Because
(a) $f(x) = e^{cx}$ is convex,
(b)

$$E[Z_n - Z_{n-1}|\mathcal{F}_{n-1}] \ = \ E[Z_n|\mathcal{F}_{n-1}] - E[Z_{n-1}|\mathcal{F}_{n-1}]$$
$$= \ Z_{n-1} - Z_{n-1} = 0,$$

and

(c) $-B_n \leq Z_n - Z_{n-1} \leq -B_n + d_n.$
It follows from Lemma 3.24, with $\alpha = B_n$, $\beta = -B_n + d_n$, that

$$E[e^{c(Z_n-Z_{n-1})}|\mathcal{F}_{n-1}] \ \leq \ E\left[\frac{(-B_n + d_n)e^{-cB_n} + B_n e^{c(-B_n+d_n)}}{+d_n}\Big|\mathcal{F}_{n-1}\right]$$
$$= \ \frac{(-B_n + d_n)e^{-cB_n} + B_n e^{c(-B_n+d_n)}}{d_n},$$

where the final equality used that $B_n \in \mathcal{F}_{n-1}$. Hence, from Equation 3.7, we see that

$$E[W_n|\mathcal{F}_{n-1}] \ \leq \ W_{n-1} \frac{(-B_n + d_n)e^{-cB_n} + B_n e^{c(-B_n+d_n)}}{d_n}$$
$$\leq \ W_{n-1} e^{c^2 d_n^2/8},$$

where the final inequality used Lemma 3.25 (with $p = B_n/d_n$, $t = cd_n$). Taking expectations gives

$$E[W_n] \leq E[W_{n-1}]e^{c^2 d_n^2/8}.$$

Using that $E[W_0] = 1$ yields, upon iterating this inequality,

$$E[W_n] \leq \prod_{i=1}^{n} e^{c^2 d_i^2/8} = e^{c^2 \sum_{i=1}^{n} d_i^2/8}.$$

Therefore, from Equation 3.6, we obtain that for any $c > 0$

$$P(Z_n \geq a) \leq \exp\left(-ca + c^2 \sum_{i=1}^{n} d_i^2/8\right).$$

Letting $c = 4a/\sum_{i=1}^{n} d_i^2$, which is the value of c that minimizes $-ca + c^2 \sum_{i=1}^{n} d_i^2/8$, gives

$$P(Z_n \geq a) \leq e^{-2a^2/\sum_{i=1}^{n} d_i^2}.$$

Parts (a) and (b) of the Hoeffding–Azuma inequality now follow from applying the preceding, first to the zero mean martingale $\{Z_n - \mu\}$ and second to the zero mean martingale $\{\mu - Z_n\}$. ∎

Example 3.27 Let $X_i, i \geq 1$ be independent Bernoulli random variables with means $p_i, i = 1, \ldots, n$. Then

$$Z_n = \sum_{i=1}^{n}(X_i - p_i) = S_n - \sum_{i=1}^{n} p_i, \ n \geq 0$$

is a martingale with mean zero. Because $Z_n - Z_{n-1} = X_n - p$, we see that

$$-p \leq Z_n - Z_{n-1} \leq 1 - p.$$

Thus, by the Hoeffding–Azuma inequality ($B_n = p, d_n = 1$), we see that for $a > 0$

$$P\left(S_n - \sum_{i=1}^{n} p_i \geq a\right) \leq e^{-2a^2/n}$$

$$P\left(S_n - \sum_{i=1}^{n} p_i \leq -a\right) \leq e^{-2a^2/n}.$$

The preceding inequalities are often called *Chernoff bounds*. ∎

The Hoeffding–Azuma inequality is often applied to a Doob type martingale. The following corollary is often used.

Corollary 3.28 *Let h be such that if the vectors $\mathbf{x} = (x_1, \ldots, x_n)$ and $\mathbf{y} = (y_1, \ldots, y_n)$ differ in at most one coordinate (that is, for some k, $x_i = y_i$ for all $i \neq k$) then*

$$|h(\mathbf{x}) - h(\mathbf{y})| \leq 1.$$

Then, for a vector of independent random variables $\mathbf{X} = (X_1, \ldots, X_n)$, and $a > 0$

$$P(h(\mathbf{X}) - E[h(\mathbf{X})] \geq a) \leq e^{-2a^2/n}$$
$$P(h(\mathbf{X}) - E[h(\mathbf{X})] \leq -a) \leq e^{-2a^2/n}.$$

Proof Let $Z_0 = E[h(\mathbf{X})]$, and $Z_i = E[h(\mathbf{X})|\sigma(X_1, \ldots, X_i)]$, for $i = 1, \ldots, n$. Then $Z_0, \ldots, Z_n$ is a martingale with respect to the filtration $\sigma(X_1, \ldots, X_i), i = 1, \ldots, n$. Now,

$$\begin{aligned}
Z_i - Z_{i-1} &= E[h(\mathbf{X})|X_1, \ldots, X_i)] - E[h(\mathbf{X})|X_1, \ldots, X_{i-1}] \\
&\leq \sup_x \{E[h(\mathbf{X})|X_1, \ldots, X_{i-1}, X_i = x] \\
&\quad - E[h(\mathbf{X})|X_1, \ldots, X_{i-1}]\}.
\end{aligned}$$

Similarly,

$$\begin{aligned}
Z_i - Z_{i-1} &\geq \inf_y \{E[h(\mathbf{X})|X_1, \ldots, X_{i-1}, X_i = y] \\
&\quad - E[h(\mathbf{X})|X_1, \ldots, X_{i-1}]\}.
\end{aligned}$$

Hence, letting

$$-B_i = \inf_y \{E[h(\mathbf{X})|X_1, \ldots, X_{i-1}, X_i = y] - E[h(\mathbf{X})|X_1, \ldots, X_{i-1}]\}$$

and $d_i = 1$, the result will follow from the Hoeffding–Azuma inequality if we can show that

$$\begin{aligned}
&\sup_x \{E[h(\mathbf{X})|X_1, \ldots, X_{i-1}, X_i = x]\} \\
&\quad - \inf_y \{E[h(\mathbf{X})|X_1, \ldots, X_{i-1}, X_i = y]\} \leq 1.
\end{aligned}$$

However, with $\mathbf{X}_{i-1} = (X_1, \ldots, X_{i-1})$, the left-hand side of the preceding can be written as

$$\sup_{x,y}\{E[h(\mathbf{X})|X_1,\ldots,X_{i-1},X_i=x]$$

$$- E[h(\mathbf{X})|X_1,\ldots,X_{i-1},X_i=y]\}$$

$$= \sup_{x,y}\{E[h(X_1,\ldots,X_{i-1},x,X_{i+1},\ldots,X_n)|\mathbf{X}_{i-1}]$$

$$- E[h(X_1,\ldots,X_{i-1},y,X_{i+1},\ldots,X_n)|\mathbf{X}_{i-1}]\}$$

$$= \sup_{x,y}\{E[h(X_1,\ldots,X_{i-1},x,X_{i+1},\ldots,X_n)$$

$$- E[h(X_1,\ldots,X_{i-1},y,X_{i+1},\ldots,X_n)|\mathbf{X}_{i-1}]\}$$

$$\leq 1,$$

and the proof is complete. ∎

Example 3.29 Let $X_1,X_2,\ldots,X_n$ be iid discrete random variables, with $P(X_i=j)=p_j$. With N equal to the number of times the pattern $3,4,5,6$, appears in the sequence $X_1,X_2,\ldots,X_n$, obtain bounds on the tail probability of N.

Solution   First note that

$$E[N] = \sum_{i=1}^{n-3} E\left[I_{\{\text{pattern begins at position } i\}}\right]$$

$$= (n-3)p_3p_4p_5p_6.$$

With $h(x_1,\ldots,x_n)$ equal to the number of times the pattern $3,4,5,6$ appears when $X_i=x_i, i=1,\ldots,n$, it is easy to see that h satisfies the condition of Corollary 3.28. Hence, for $a>0$

$$P(N-(n-3)p_3p_4p_5p_6 \geq a) \leq e^{-2a^2/(n-3)}$$

$$P(N-(n-3)p_3p_4p_5p_6 \leq -a) \leq e^{-2a^2/(n-3)}.\qquad\blacksquare$$

Example 3.30 Suppose that n balls are to be placed in m urns, with each ball independently going into urn i with probability p_i, $\sum_{i=1}^{m} p_i = 1$. Find bounds on the tail probability of Y_k, equal to the number of urns that receive exactly k balls.

Solution   First note that

$$E[Y_k] = E\left[\sum_{i=1}^{m} I_{\{\text{urn } i \text{ has exactly } k \text{ balls}\}}\right]$$

$$= \sum_{i=1}^{m} \binom{n}{i} p_i^k (1-p_i)^{n-k}.$$

Let X_j denote the urn in which ball j is put, where $j = 1, \ldots, n$. Also, let $h_k(x_1, \ldots, x_n)$ denote the number of urns that receive exactly k balls when $X_i = x_i, i = 1, \ldots, n$, and note that $Y_k = h_k(X_1, \ldots, X_n)$. When $k = 0$, it is easy to see that h_0 satisfies the condition that if $\mathbf{x}$ and $\mathbf{y}$ differ in at most one coordinate, then $|h_0(\mathbf{x}) - h_0(\mathbf{y})| \leq 1$. Therefore, from Corollary 3.28 we obtain, for $a > 0$, that

$$P\left(Y_0 - \sum_{i=1}^m (1-p_i)^n \geq a\right) \leq e^{-2a^2/n}$$

$$P\left(Y_0 - \sum_{i=1}^m (1-p_i)^n \leq -a\right) \leq e^{-2a^2/n}.$$

For $0 < k < n$, it is no longer true that if $\mathbf{x}$ and $\mathbf{y}$ differ in at most one coordinate, then $|h_k(\mathbf{x}) - h_k(\mathbf{y})| \leq 1$. This is because the one different value could result in one of the vectors having one less and the other having one more urn with k balls than would have resulted if that coordinate was not included. Thus, if $\mathbf{x}$ and $\mathbf{y}$ differ in at most one coordinate, then

$$|h_k(\mathbf{x}) - h_k(\mathbf{y})| \leq 2,$$

showing that $h_k^*(\mathbf{x}) = h_k(\mathbf{x})/2$ satisfies the condition of Corollary 3.28. Because

$$P(Y_k - E[Y_k] \geq a) = P(h_k^*(\mathbf{X}) - E[h_k^*(\mathbf{X})] \geq a/2),$$

we obtain, for $a > 0, 0 < k < n$, that

$$P\left(Y_k - \sum_{i=1}^m \binom{n}{i} p_i^k (1-p_i)^{n-k} \geq a\right) \leq e^{-a^2/2n}$$

$$P\left(Y_k - \sum_{i=1}^m \binom{n}{i} p_i^k (1-p_i)^{n-k} \leq -a\right) \leq e^{-a^2/2n}.$$

Of course,

$$P(Y_n = 1) = \sum_{i=1}^m p_i^n = 1 - P(Y_n = 0). \quad \blacksquare$$

3.6 Submartingales, Supermartingales, and a Convergence Theorem

Submartingales model superfair games, whereas supermartingales model subfair ones.

Definition 3.31 *The sequence of random variables Z_n, $n \geq 1$, is said to be a submartingale for the filtration $\mathcal{F}_n$ if*

(a) $E[|Z_n|] < \infty$

(b) Z_n is adapted to $\mathcal{F}_n$

(c) $E[Z_{n+1}|\mathcal{F}_n] \geq Z_n$

If Part (c) is replaced by $E[Z_{n+1}|\mathcal{F}_n] \leq Z_n$, then $Z_n, n \geq 1$, is said to be supermartingale.

It follows from the tower property that

$$E[Z_{n+1}] \geq E[Z_n]$$

for a submartingale, with the inequality reversed for a supermartingale. (Of course, if $Z_n, n \geq 1$, is a submartingale, then $-Z_n, n \geq 1$, is a supermartingale and vice-versa.)

The analogs of the martingale stopping theorem remain valid for submartingales and supermartingales. We leave the proof of the following theorem as an exercise.

Theorem 3.32 *If N is a stopping time for the filtration $\mathcal{F}_n$, then*

$$E[Z_N] \geq E[Z_1] \quad \text{for a submartingale}$$
$$E[Z_N] \leq E[Z_1] \quad \text{for a supermartingale}$$

provided that any of the sufficient conditions of Theorem 3.14 hold.

One of the most useful results about submartingales is the Kolmogorov inequality. Before presenting it, we need a couple of lemmas.

Lemma 3.33 *If $Z_n, n \geq 1$, is a submartingale for the filtration $\mathcal{F}_n$, and N is a stopping time for this filtration such that $P(N \leq n) = 1$, then*

$$E[Z_1] \leq E[Z_N] \leq E[Z_n].$$

Proof Because N is bounded, it follows from the submartingale stopping theorem that $E[Z_N] \geq E[Z_1]$. Now,

$$E[Z_n|\mathcal{F}_k, N = k] = E[Z_n|\mathcal{F}_k] \geq Z_k = Z_N.$$

Taking expectations of this inequality completes the proof. ∎

Lemma 3.34 *If $Z_n, n \geq 1$, is a martingale with respect to the filtration $\mathcal{F}_n, n \geq 1$, and f is a convex function for which $E[|f(Z_n)|] < \infty$, then $f(Z_n), n \geq 1$, is a submartingale with respect to the filtration $\mathcal{F}_n, n \geq 1$.*

Proof

$$E[f(Z_{n+1})|\mathcal{F}_n] \geq f(E[Z_{n+1}|\mathcal{F}_n]) \qquad \text{by Jensen's inequality}$$
$$= f(Z_n).$$

∎

Theorem 3.35 *Kolmogorov's inequality for submartingales. Suppose $Z_n, n \geq 1$, is a nonnegative submartingale, then for $a > 0$*

$$P(\max\{Z_1, \ldots, Z_n\} \geq a) \leq E[Z_n]/a.$$

Proof Let N be the smallest i, $i \leq n$ such that $Z_i \geq a$, and let it equal n if $Z_i < a$ for all $i = 1, \ldots, n$. Then

$$P(\max\{Z_1, \ldots, Z_n\} \geq a) = P(Z_N \geq a)$$
$$\leq E[Z_N]/a \qquad \text{by Markov's inequality}$$
$$\leq E[Z_n]/a \qquad \text{since } N \leq n. \qquad ∎$$

Corollary 3.36 *If $Z_n, n \geq 1$, is a martingale, then for $a > 0$*

$$P(\max\{|Z_1|, \ldots, |Z_n|\} \geq a) \leq \min(E[|Z_n|]/a, \ E[Z_n^2]/a^2).$$

Proof Noting that

$$P(\max\{|Z_1|, \ldots, |Z_n|\} \geq a) = P(\max\{Z_1^2 \ldots, Z_n^2\} \geq a^2)$$

the corollary follows from Lemma 3.34 and Kolmogorov's inequality for submartingales upon using that $f(x) = |x|$ and $f(x) = x^2$ are convex functions. ∎

Theorem 3.37 *Martingale convergence theorem. Let Z_n, $n \geq 1$, be a martingale. If there is $M < \infty$ such that*

$$E[|Z_n|] \leq M \qquad \text{for all } n,$$

then with a probability of one $\lim_{n \to \infty} Z_n$ exists and is finite.

Proof We will give a proof under the stronger condition that $E[Z_n^2]$ is bounded. Because $f(x) = x^2$ is convex, it follows from Lemma 3.34 that $Z_n^2, n \geq 1$, is a submartingale, yielding that $E[Z_n^2]$ is nondecreasing. Because $E[Z_n^2]$ is bounded, it follows that it converges; let $m < \infty$ be given by

$$m = \lim_{n \to \infty} E[Z_n^2].$$

We now argue that $\lim_{n\to\infty} Z_n$ exists and is finite by showing that, with a probability of one $Z_n, n \geq 1$, is a Cauchy sequence. That is, we will show that, with a probability of one

$$|Z_{m+k} - Z_m| \to 0 \quad \text{as } m, k \to \infty.$$

Using that $Z_{m+k} - Z_m, k \geq 1$, is a martingale, it follows that $(Z_{m+k} - Z_m)^2, k \geq 1$, is a submartingale. Thus, by Kolmogorov's inequality,

$$P(|Z_{m+k} - Z_m| > \epsilon \quad \text{for some } k \leq n)$$

$$= P\left(\max_{k=1,\ldots,n} (Z_{m+k} - Z_m)^2 > \epsilon^2\right)$$
$$\leq E[(Z_{m+n} - Z_m)^2]/\epsilon^2$$
$$= E[Z_{n+m}^2 - 2Z_m Z_{n+m} + Z_m^2]/\epsilon^2.$$

However,

$$\begin{aligned} E[Z_m Z_{n+m}] &= E[E[Z_m Z_{n+m}|Z_m]] \\ &= E[Z_m E[Z_{n+m}|Z_m]] \\ &= E[Z_m^2]. \end{aligned}$$

Therefore,

$$P(|Z_{m+k} - Z_m| > \epsilon \quad \text{for some } k \leq n) \leq (E[Z_{n+m}^2] - E[Z_m^2])/\epsilon^2.$$

Letting $n \to \infty$ now yields

$$P(|Z_{m+k} - Z_m| > \epsilon \quad \text{for some } k) \leq (m - E[Z_m^2])/\epsilon^2.$$

Thus,

$$P(|Z_{m+k} - Z_m| > \epsilon \quad \text{for some } k) \to 0 \quad \text{as } m \to \infty.$$

Therefore, with a probability of one, $Z_n, n \geq 1$, is a Cauchy sequence, and so has a finite limit. ∎

As a consequence of the martingale convergence theorem, we obtain the following.

Corollary 3.38 *If $Z_n, n \geq 1$, is a nonnegative martingale, then with a probability of one, $\lim_{n\to\infty} Z_n$ exists and is finite.*

Proof Because Z_n is nonnegative,

$$E[|Z_n|] = E[Z_n] = E[Z_1] < \infty. \qquad \blacksquare$$

Example 3.39 A *branching process* follows the size of a population over succeeding generations. It supposes that, independent of what occurred in prior generations, each individual in generation n independently has j offspring with probability $p_j, j \geq 0$. The offspring of individuals of generation n then make up generation $n+1$. Let X_n denote the number of individuals in generation n. Assuming that $m = \sum_j jp_j$, the mean number of offspring of an individual, is finite it is easy to verify that $Z_n = X_n/m^n, n \geq 0$, is a martingale. Because it is nonnegative, the preceding corollary implies that $\lim_n X_n/m^n$ exists and is finite. But this implies, when $m < 1$, that $\lim_n X_n = 0$ or, equivalently, that $X_n = 0$ for all n sufficiently large. When $m > 1$, the implication is that the generation size either becomes zero or converges to infinity at an exponential rate. ∎

3.7 Exercises

1. For $\mathcal{F} = \{\phi, \Omega\}$, show that $E[X|\mathcal{F}] = E[X]$.

2. Give the proof of Proposition 3.2 when X and Y are jointly continuous.

3. If $E[|X_i|] < \infty, i = 1, \ldots, n$, show that

$$E\left[\sum_{i=1}^{n} X_i \Big| \mathcal{F}\right] = \sum_{i=1}^{n} E[X_i|\mathcal{F}].$$

4. Prove that if f is a convex function, then

$$E[f(X)|\mathcal{F}] \geq f(E[X|\mathcal{F}])$$

   provided the expectations exist.

5. Let $X_1, X_2, \ldots$ be independent random variables with mean one. Show that $Z_n = \prod_{i=1}^{n} X_i, n \geq 1$, is a martingale.

6. If $E[X_{n+1}|X_1, \ldots, X_n] = a_n X_n + b_n$ for constants $a_n, b_n, n \geq 0$, find constants A_n, B_n so that $Z_n = A_n X_n + B_n, n \geq 0$, is a martingale with respect to the filtration $\sigma(X_0, \ldots, X_n)$.

7. Consider a population of individuals as it evolves over time, and suppose that, independent of what occurred in prior generations, each individual in generation n independently has j offspring with probability $p_j, j \geq 0$. The offspring of individuals of generation n then make up generation $n+1$. Assume that $m = \sum_j jp_j < \infty$. Let X_n denote the number of individuals in generation n, and define a martingale related to $X_n, n \geq 0$. The process $X_n, n \geq 0$ is called a branching process.

8. Suppose $X_1, X_2, \ldots$ are iid random variables with mean zero and finite variance σ^2. If T is a stopping time with finite mean, show that

$$\mathrm{Var}\left(\sum_{i=1}^{T} X_i\right) = \sigma^2 E(T).$$

9. Suppose $X_1, X_2, \ldots$ are iid mean zero random variables that each take value $+1$ with probability $1/2$ and take value -1 with probability $1/2$. Let $S_n = \sum_{i=1}^{n} X_i$. Which of the following (a)–(c) are stopping times? Compute $E[T_i]$ for the T_i that are stopping times.
   (a) $T_1 = \min\{i \geq 5 : S_i = S_{i-5} + 5\}$.
   (b) $T_2 = T_1 - 5$.
   (c) $T_3 = T_2 + 10$.

10. Consider a sequence of independent flips of a coin, and let P_h denote the probability of a head on any toss. Let A be the hypothesis that $P_h = a$, and let B be the hypothesis that $P_h = b$, for given values $0 < a, b < 1$. Let X_i be the outcome of flip i, and set

$$Z_n = \frac{P(X_1, \ldots, X_n | A)}{P(X_1, \ldots, X_n | B)}.$$

    If $P_h = b$, show that $Z_n, n \geq 1$, is a martingale having mean one.

11. Let $Z_n, n \geq 0$ be a martingale with $Z_0 = 0$. Show that

$$E[Z_n^2] = \sum_{i=1}^{n} E[(Z_i - Z_{i-1})^2].$$

12. Consider an individual who at each stage, independently of past movements, moves to the right with probability p or to the left with probability $1 - p$. Assuming that $p > 1/2$, find the expected number of stages it takes the person to move i positions to the right from where they started.

13. In Example 3.19 obtain bounds on p when $\theta < 0$.

14. Use Wald's equation to approximate the expected time it takes a random walk to either become as large as a or as small as $-b$, for positive a and b. Give the exact expression if a and b are integers, and at each stage the random walk either moves up one with probability p or moves down one with probability $1 - p$.

15. Consider a branching process that starts with a single individual. Let π denote the probability this process eventually dies out. With X_n denoting the number of individuals in generation n, argue that $\pi^{X_n}, n \geq 0$, is a martingale.

16. Given $X_1, X_2, \ldots$, let $S_n = \sum_{i=1}^{n} X_i$ and $\mathcal{F}_n = \sigma(X_1, \ldots X_n)$. Suppose for all n $E|S_n| < \infty$ and $E[S_{n+1}|\mathcal{F}_n] = S_n$. Show $E[X_i X_j] = 0$ if $i \neq j$.

17. Suppose n random points are chosen in a circle having a diameter equal to one, and let X be the length of the shortest path connecting all of them. For $a > 0$, bound $P(X - E[X] \geq a)$.

18. Let $X_1, X_2, \ldots, X_n$ be iid discrete random variables, with $P(X_i = j) = p_j$. Obtain bounds on the tail probability of the number of times the pattern $0, 0, 0, 0$ appears in the sequence.

19. Repeat Example 3.29, but now assume that the X_i are independent but not identically distributed. Let $P_{i,j} = P(X_i = j)$.

20. Let $Z_n, n \geq 0$, be a martingale with mean $Z_0 = 0$, and let $v_j, j \geq 0$, be a sequence of nondecreasing constants with $v_0 = 0$. Prove the *Kolmogorov-Hajek-Renyi inequality*:

$$P(|Z_j| \leq v_j, \text{ for all } j = 1, \ldots, n) \geq 1 - \sum_{j=1}^{n} E[(Z_j - Z_{j-1})^2]/v_j^2.$$

21. Consider a gambler who plays at a fair casino. Suppose that the casino does not give any credit, so the gambler must quit when their fortune is zero. Suppose further that on each bet made at least one is either won or lost. Argue that, with a probability of one, a gambler who wants to play forever will eventually go broke.

22. What is the implication of the martingale convergence theorem to the scenario of Exercise 10?

23. Three gamblers each start with a, b, and c chips, respectively. In each round of a game, a gambler is selected uniformly at random to give up a chip, and one of the other gamblers is selected uniformly at random to receive that chip. The game ends when there are only two players remaining with chips. Let X_n, Y_n, and Z_n respectively denote the number of chips the three players have after round n, so $(X_0, Y_0, Z_0) = (a, b, c)$.
    (a) Compute $E[X_{n+1}Y_{n+1}Z_{n+1} \mid (X_n, Y_n, Z_n) = (x, y, z)]$.
    (b) Show that $M_n = X_n Y_n Z_n + n(a + b + c)/3$ is a martingale.
    (c) Use the preceding to compute the expected length of the game.

24. In the ante one game of Example 3.22, find the expected number of games played by player 1.

25. Consider the ante one game, but now suppose there are r players with each player initially having fortune k. Suppose further that each

player has a value, with v_i being the value of player i. As before, suppose that a player is alive if their fortune is positive and that at the beginning of a round each alive player puts one into the pot. But with A being the set of alive players, now suppose the pot is won by $i \in A$ with probability $\frac{v_i}{\sum_{j \in A} v_j}$.

(a) With $X_i(n)$ being player i's fortune after game n, find $E[X_i(k)]$.

(b) Let P_i be the probability that player i eventually has fortune kr. If $v_1 = \max_i v_i$, show that $P_1 \geq \frac{v_1}{\sum_{i=1}^{r} v_i}$.

26. Suppose a gambler bets \$1 that a fair coin will come up heads. If it comes up heads, they win their bet and stop; if it comes up tails, they lose their bet, and in subsequent flips, they will bet all their losses so far plus \$1 that the next flip will be heads. This repeats until the gambler eventually wins and has a \$1 net profit. Because all bets are fair, why doesn't the martingale stopping theorem apply to show that the gambler is not expected to come out ahead?

27. Suppose a gambler wins \$1 each time a flipped coin comes up heads, and loses \$1 each time it comes up tails. Suppose the coin is flipped until the gambler eventually ends up with a \$1 profit. Because all bets are fair, why doesn't the martingale stopping theorem apply to show that the gambler is not expected to come out ahead? It can be shown using Markov chain theory that, with a probability of one, the gambler will eventually be up \$1.

4

Bounding Probabilities and Expectations

4.1 Introduction

In this chapter, we develop some approaches for bounding expectations and probabilities. We start in Section 4.2 with Jensen's inequality, which bounds the expected value of a convex function of a random variable. In Section 4.3, we develop the importance sampling identity and show how it can be used to yield bounds on tail probabilities. A specialization of this method results in the Chernoff bound, which is developed in Section 4.4. Section 4.5 deals with the second moment and the conditional expectation inequalities, which lower bound the probability that at least one of a given number of events occurs. Section 4.6 develops the min-max identity and uses it to obtain bounds on the maximum of a set of random variables. Finally, in Section 4.7 we introduce some general stochastic order relations and explore their consequences.

4.2 Jensen's Inequality

Jensen's inequality yields a lower bound on the expected value of a convex function of a random variable.

Proposition 4.1 *If f is a convex function, then*

$$E[f(X)] \geq f(E[X])$$

provided the expectations exist.

Proof We give a proof under the assumption that f has a Taylor series expansion. Expanding f about the value $\mu = E[X]$, and using the Taylor series expansion with a remainder term, yields that for some a

$$
\begin{aligned}
f(x) &= f(\mu) + f'(\mu)(x - \mu) + f''(a)(x - \mu)^2/2 \\
&\geq f(\mu) + f'(\mu)(x - \mu),
\end{aligned}
$$

where the preceding used that $f''(a) \geq 0$ by convexity. Hence,

$$
f(X) \geq f(\mu) + f'(\mu)(X - \mu).
$$

Taking expectations yields the result.　■

Remark 4.2 If

$$
P(X = x_1) = \lambda = 1 - P(X = x_2),
$$

then Jensen's inequality implies that for a convex function f

$$
\lambda f(x_1) + (1 - \lambda)f(x_2) \geq f(\lambda x_1 + (1 - \lambda)x_2),
$$

which is the definition of a convex function. Thus, Jensen's inequality can be thought of as extending the defining equation of convexity from random variables that take on only two possible values to arbitrary random variables.

4.3　Probability Bounds via the Importance Sampling Identity

Let f and g be probability density (or probability mass) functions; let h be an arbitrary function, and suppose that $g(x) = 0$ implies that $f(x)h(x) = 0$. The following is known as the importance sampling identity.

Proposition 4.3 *The importance sampling identity is*

$$
E_f[h(X)] = E_g\left[\frac{h(X)f(X)}{g(X)}\right],
$$

where the subscript on the expectation indicates the density (or mass function) of the random variable X.

Proof We give the proof when f and g are density functions:

$$
\begin{aligned}
E_f[h(X)] &= \int_{-\infty}^{\infty} h(x)\,f(x)\,dx \\
&= \int_{-\infty}^{\infty} \frac{h(x)\,f(x)}{g(x)}\,g(x)\,dx \\
&= E_g\left[\frac{h(X)f(X)}{g(X)}\right].　■
\end{aligned}
$$

The importance sampling identity yields the following useful corollary concerning the tail probability of a random variable.

Corollary 4.4

$$P_f(X > c) = E_g\left[\frac{f(X)}{g(X)}\Big| X > c\right] P_g(X > c).$$

Proof

$$\begin{aligned}
P_f(X > c) &= E_f[I_{\{X>c\}}] \\
&= E_g\left[\frac{I_{\{X>c\}}f(X)}{g(X)}\right] \\
&= E_g\left[\frac{I_{\{X>c\}}f(X)}{g(X)}\Big| X > c\right] P_g(X > c) \\
&= E_g\left[\frac{f(X)}{g(X)}\Big| X > c\right] P_g(X > c).
\end{aligned}$$

∎

Example 4.5 *Bounding standard normal tail probabilities.* Let f be the standard normal density function

$$f(x) = \frac{1}{\sqrt{2\pi}} e^{-x^2/2}, \quad -\infty < x < \infty.$$

For $c > 0$, consider $P_f(X > c)$, the probability that a standard normal random variable exceeds c. With

$$g(x) = ce^{-cx}, \quad x > 0,$$

we obtain from Corollary 4.4

$$\begin{aligned}
P_f(X > c) &= \frac{e^{-c^2}}{c\sqrt{2\pi}} E_g[e^{-X^2/2}e^{cX}|X > c] \\
&= \frac{e^{-c^2}}{c\sqrt{2\pi}} E_g[e^{-(X+c)^2/2}e^{c(X+c)}],
\end{aligned}$$

where the first equality used that $P_g(X > c) = e^{-c^2}$ and the second the lack of memory property of exponential random variables to conclude that the conditional distribution of an exponential random variable X given that it exceeds c is the unconditional distribution of $X + c$. Thus the preceding yields

$$P_f(X > c) = \frac{e^{-c^2/2}}{c\sqrt{2\pi}} E_g[e^{-X^2/2}]. \tag{4.1}$$

Noting that, for $x > 0$,

$$1 - x < e^{-x} < 1 - x + x^2/2,$$

we see that

$$1 - X^2/2 < e^{-X^2/2} < 1 - X^2/2 + X^4/8.$$

Using that $E[X^2] = 2/c^2$ and $E[X^4] = 24/c^4$ when X is exponential with rate c, the preceding inequality yields

$$1 - 1/c^2 < E_g[e^{-X^2/2}] < 1 - 1/c^2 + 3/c^4.$$

Consequently, using Equation 4.1, we obtain

$$(1 - 1/c^2) \frac{e^{-c^2/2}}{c\sqrt{2\pi}} < P_f(X > c) < (1 - 1/c^2 + 3/c^4) \frac{e^{-c^2/2}}{c\sqrt{2\pi}}. \qquad \blacksquare \quad (4.2)$$

Our next example uses the importance sampling identity to bound the probability that successive sums of a sequence of iid normal random variables with a negative mean ever cross some specified positive number.

Example 4.6 Let $X_1, X_2, \ldots$ be a sequence of iid normal random variables with mean $\mu < 0$ and variance one. Let $S_k = \sum_{i=1}^k X_i$ and, for a fixed $A > 0$, consider

$$p = P(S_k > A \text{ for some } k).$$

Let $f_k(\mathbf{x}_k) = f_k(x_1, \ldots, x_k)$ be the joint density function of $\mathbf{X}_k = (X_1, \ldots, X_k)$. That is,

$$f_k(\mathbf{x}_k) = (2\pi)^{-k/2} e^{-\sum_{i=1}^k (x_i - \mu)^2/2}.$$

Also, let g_k be the joint density of k iid normal random variables with mean $-\mu$ and variance one. That is,

$$g_k(\mathbf{x}_k) = (2\pi)^{-k/2} e^{-\sum_{i=1}^k (x_i + \mu)^2/2}.$$

Note that

$$\frac{f_k(\mathbf{x}_k)}{g_k(\mathbf{x}_k)} = e^{2\mu \sum_{i=1}^k x_i}.$$

With

$$R_k = \left\{ (x_1, \ldots, x_k) : \sum_{i=1}^j x_i \le A, \, j < k, \, \sum_{i=1}^k x_i > A \right\},$$

we have

$$p = \sum_{k=1}^{\infty} P(\mathbf{X}_k \in R_k)$$

$$= \sum_{k=1}^{\infty} E_{f_k}[I_{\{\mathbf{X}_k \in R_k\}}]$$

$$= \sum_{k=1}^{\infty} E_{g_k}\left[\frac{I_{\{\mathbf{X}_k \in R_k\}} f_k(\mathbf{X}_k)}{g_k(\mathbf{X}_k)}\right]$$

$$= \sum_{k=1}^{\infty} E_{g_k}[I_{\{\mathbf{X}_k \in R_k\}} e^{2\mu S_k}].$$

Now, if $\mathbf{X}_k \in R_k$ then $S_k > A$, implying, because $\mu < 0$, that $e^{2\mu S_k} < e^{2\mu A}$. Because this implies that

$$I_{\{\mathbf{X}_k \in R_k\}} e^{2\mu S_k} \le I_{\{\mathbf{X}_k \in R_k\}} e^{2\mu A},$$

we obtain from the preceding that

$$p \le \sum_{k=1}^{\infty} E_{g_k}[I_{\{\mathbf{X}_k \in R_k\}} e^{2\mu A}]$$

$$= e^{2\mu A} \sum_{k=1}^{\infty} E_{g_k}[I_{\{\mathbf{X}_k \in R_k\}}].$$

Now, if $Y_i, i \ge 1$, is a sequence of independent normal random variables with mean $-\mu$ and variance 1, then

$$E_{g_k}[I_{\{\mathbf{X}_k \in R_k\}}] = P\left(\sum_{i=1}^{j} Y_i \le A, ,j < k, \sum_{i=1}^{k} Y_i > A\right).$$

Therefore, from the preceding

$$p \le e^{2\mu A} \sum_{k=1}^{\infty} P\left(\sum_{i=1}^{j} Y_i \le A, ,j < k, \sum_{i=1}^{k} Y_i > A\right)$$

$$= e^{2\mu A} P\left(\sum_{i=1}^{k} Y_i > A \text{ for some } k\right)$$

$$= e^{2\mu A},$$

where the final equality follows from the strong law of large numbers because $\lim_{n \to \infty} \sum_{i=1}^{n} Y_i/n = -\mu > 0$, implies $P(\lim_{n \to \infty} \sum_{i=1}^{n} Y_i = \infty) = 1$, and thus $P(\sum_{i=1}^{k} Y_i > A \text{ for some } k) = 1$.

The bound

$$p = P(S_k > A \text{ for some } k) \leq e^{2\mu A}$$

is not useful when A is a small nonnegative number. In this case, we should condition on X_1 and then apply the preceding inequality. With Φ being the standard normal distribution function, this yields

$$
\begin{aligned}
p &= \int_{-\infty}^{\infty} P(S_k > A \text{ for some } k | X_1 = x) \frac{1}{\sqrt{2\pi}} e^{-(x-\mu)^2/2} dx \\
&= \int_{-\infty}^{A} P(S_k > A \text{ for some } k | X_1 = x) \frac{1}{\sqrt{2\pi}} e^{-(x-\mu)^2/2} dx \\
&\quad + P(X_1 > A) \\
&\leq \frac{1}{\sqrt{2\pi}} \int_{-\infty}^{A} e^{2\mu(A-x)} e^{-(x-\mu)^2/2} dx + 1 - \Phi(A - \mu) \\
&= e^{2\mu A} \frac{1}{\sqrt{2\pi}} \int_{-\infty}^{A} e^{-(x+\mu)^2/2} dx + 1 - \Phi(A - \mu) \\
&= e^{2\mu A} \Phi(A + \mu) + 1 - \Phi(A - \mu).
\end{aligned}
$$

Thus, for instance

$$P(S_k > 0 \text{ for some } k) \leq \Phi(\mu) + 1 - \Phi(-\mu) = 2\Phi(\mu).$$

4.4   Chernoff Bounds

Suppose that X has probability density (or probability mass) function $f(x)$. For $t > 0$, let

$$g(x) = \frac{e^{tx} f(x)}{M(t)},$$

where $M(t) = E_f[e^{tX}]$ is the moment generating function of X. Corollary 4.4 yields, for $c > 0$, that

$$
\begin{aligned}
P_f(X \geq c) &= E_g[M(t)e^{-tX} | X \geq c] P_g(X \geq c) \\
&\leq E_g[M(t)e^{-tX} | X \geq c] \\
&\leq M(t) e^{-tc}.
\end{aligned}
$$

Because the preceding holds for all $t > 0$, we can conclude that

$$P_f(X \geq c) \leq \inf_{t>0} M(t) e^{-tc}. \qquad (4.3)$$

The inequality (Equation 4.3) is called the *Chernoff bound*.

Rather than choosing the value of t so as to obtain the best bound, it is often convenient to work with bounds that are more analytically tractable. The following inequality can be used to simplify the Chernoff bound for a sum of independent Bernoulli random variables.

Lemma 4.7   *For $0 \leq p \leq 1$,*

$$pe^{t(1-p)} + (1-p)e^{-tp} \leq e^{t^2/8}.$$

The proof of Lemma 4.7 was given in Lemma 3.25.

Corollary 4.8   *Let $X_1, \ldots, X_n$ be independent Bernoulli random variables, and set $W = \sum_{i=1}^n X_i$. Then, for any $c > 0$,*

$$P(W - E[W] \geq c) \leq e^{-2c^2/n} \tag{4.4}$$

$$P(W - E[W] \leq -c) \leq e^{-2c^2/n}. \tag{4.5}$$

Proof For $c > 0$, $t > 0$,

$$
\begin{aligned}
P(W - E[W] \geq c) &= P(e^{t(W-E[W])} \geq e^{tc}) \\
&= e^{-tc} E[e^{t(W-E[W])}] \quad \text{by the Markov inequality} \\
&= e^{-tc} E\left[\exp\left\{ \sum_{i=1}^n t(X_i - E[X_i]) \right\} \right] \\
&= e^{-tc} E\left[\prod_{i=1}^n e^{t(X_i - E[X_i])} \right] \\
&= e^{-tc} \prod_{i=1}^n E[e^{t(X_i - E[X_i])}].
\end{aligned}
$$

However, if Y is Bernoulli with parameter p, then

$$E[e^{t(Y-E[Y])}] = pe^{t(1-p)} + (1-p)e^{-tp} \leq e^{t^2/8},$$

where the inequality follows from Lemma 4.7. Therefore,

$$P(W - E[W] \geq c) \leq e^{-tc} e^{nt^2/8}.$$

Letting $t = 4c/n$ yields the inequality in Equation 4.4.

The proof of the inequality in Equation 4.5 is obtained by writing it as

$$P(E[W] - W \geq c) \leq e^{-2c^2/n}$$

and using an analogous argument. ∎

Example 4.9 Suppose that an entity contains $n + m$ cells, of which cells numbered $1, \ldots, n$ are target cells, whereas cells $n + 1, \ldots, n + m$ are normal cells. Each of these $n + m$ cells has an associated weight, with w_i being the weight of cell i. Suppose that the cells are destroyed one at a time in a random order such that if S is the current set of surviving cells, then the next cell destroyed is $i, i \in S$, with probability $w_i / \sum_{j \in S} w_j$. In other words, the probability that a specified surviving cell is the next one destroyed is equal to its weight divided by the weights of all still surviving cells. Suppose that each of the n target cells has weight one, whereas each of the m normal cells has weight w. For a specified value of α, $0 < \alpha < 1$, let N_α equal the number of normal cells that are still alive at the moment when the number of surviving target cells first falls below αn. We will now show that as $n, m \to \infty$, the probability mass function of N_α becomes concentrated about the value $m\alpha^w$.

Theorem 4.10 *For any $\epsilon > 0$, as $n \to \infty$ and $m \to \infty$,*

$$P((1 - \epsilon)m\alpha^w \leq N_\alpha \leq (1 + \epsilon)m\alpha^w) \to 1.$$

Proof To prove the result, it is convenient to first formulate an equivalent continuous time model that results in the times at which the $n + m$ cells are killed being independent random variables. To do so, let $X_1, \ldots, X_{n+m}$ be independent exponential random variables, with X_i having weight w_i, $i = 1, \ldots, n + m$. Note that X_i will be the smallest of these exponentials with probability $w_i / \sum_j w_j$; further, given that X_i is the smallest, $X_r, r \neq i$, will be the second smallest with probability $w_r / \sum_{j \neq i} w_j$; further, given that X_i and X_r are, respectively, the first and second smallest, $X_s, s \neq i, r$, will be the next smallest with probability $w_s / \sum_{j \neq i, r} w_j$; and so on. Consequently, if we imagine that cell i is killed at time X_i, then the order in which the $n + m$ cells are killed has the same distribution as the order in which they are killed in the original model. So let us suppose that cell i is killed at time $X_i, i \geq 1$.

Now let τ_α denote the time at which the number of surviving target cells first falls below $n\alpha$. Also, let $N(t)$ denote the number of normal cells that are still alive at time t, so $N_\alpha = N(\tau_\alpha)$. We will first show that

$$P(N(\tau_\alpha) \leq (1 + \epsilon)m\alpha^w) \to 1 \quad \text{as } n, m \to \infty. \tag{4.6}$$

To prove the preceding, let ϵ^* be such that $0 < \epsilon^* < \epsilon$, and set $t = -\ln(\alpha(1 + \epsilon^*)^{1/w})$. We will prove Equation 4.6 by showing that as n and m approach ∞,

(a) $P(\tau_\alpha > t) \to 1$,

and

(b) $P(N(t) \leq (1 + \epsilon)m\alpha^w) \to 1$.

Because the events $\tau_\alpha > t$ and $N(t) \leq (1 + \epsilon)m\alpha^w$ together imply that $N(\tau_\alpha) \leq N(t) \leq (1 + \epsilon)m\alpha^w$, the result (Equation 4.6) will be established.

To prove Part (a), note that the number, call it Y, of surviving target cells by time t is a binomial random variable with parameters n and $e^{-t} = \alpha(1 + \epsilon^*)^{1/w}$. Hence, with $a = n\alpha[(1 + \epsilon^*)^{1/w} - 1]$, we have

$$P(\tau_\alpha \leq t) = P(Y \leq n\alpha\} = P(Y \leq ne^{-t} - a) \leq e^{-2a^2/n},$$

where the inequality follows from the Chernoff bound (Equation 4.3). This proves Part (a), because $a^2/n \to \infty$ as $n \to \infty$.

To prove Part (b), note that $N(t)$ is a binomial random variable with parameters m and $e^{-wt} = \alpha^w(1 + \epsilon^*)$. Thus, by letting $b = m\alpha^w(\epsilon - \epsilon^*)$ and again applying the Chernoff bound (Equation 4.3), we obtain

$$P(N(t) > (1 + \epsilon)m\alpha^w\} = P(N(t) > me^{-wt} + b)$$
$$\leq e^{-2b^2/m}.$$

This proves Part (b), because $b^2/m \to \infty$ as $m \to \infty$. Thus, Equation 4.6 is established.

It remains to prove that

$$P(N(\tau_\alpha) \geq (1 - \epsilon)m\alpha^w) \to 1 \quad \text{as } n, m \to \infty. \tag{4.7}$$

However, Equation 4.7 can be proven in a similar manner as Equation 4.6; a combination of these two results completes the proof of the theorem. ∎

Lemma 4.11 *Suppose $W = \sum_{i=1}^n X_i$, where $X_1, X_2, \ldots, X_n$ are indicator variables, and let I be independent of these and have a uniform distribution on the integers $1, 2, \ldots, n$. Then, for any function f,*

$$E[f(W)|X_I = 1] = \frac{E[Wf(W)]}{E[W]}.$$

Proof For any function f, we have

$$E[f(W)|X_I = 1] = E[X_I f(W)]/P(X_I = 1)$$
$$= \frac{\frac{1}{n}\sum_{i=1}^n E[X_i f(W)]}{\frac{1}{n}\sum_{i=1}^n E[X_i]}$$
$$= E[Wf(W)]/E[W].$$

∎

Proposition 4.12 *Suppose $W = \sum_{i=1}^n X_i$, where $X_1, X_2, \ldots, X_n$ are indicator variables, and let I be independent of these and have a uniform distribution on the integers $1, 2, \ldots, n$. If*

$$(W|X_I = 1) \leq_{st} W + 1, \tag{4.8}$$

then

$$P(W > x) \leq (eE[W]/x)^x.$$

Proof Let $t > 0$, and let $g(x) = e^{tx}$. By Lemma 4.11,

$$E[e^{tW}|X_I = 1] = \frac{E[We^{tW}]}{E[W]}.$$

Because $g(x)$ is an increasing function, it follows from the preceding upon using Equation 4.8 and Exercise 2a from Chapter 2 that

$$E[e^{t(W+1)}] \geq E[We^{tW}]/E[W],$$

which after letting $f(t) = E[e^{tW}]$, corresponds to

$$e^t f(t) \geq f'(t)/E[W]$$

or

$$\frac{\partial}{\partial t} \log f(t) \leq E[W]e^t$$

and

$$\log f(t) = \int_{-\infty}^{t} \frac{\partial}{\partial s}(\log f(s))ds \leq \int_{-\infty}^{t} E[W]e^s ds = E[W]e^t.$$

Using the Chernoff bound, we have

$$P(W > x) \leq e^{-xt}e^{E[W]e^t} \leq (eE[W]/x)^x,$$

where we have plugged in the minimizing value $t = \log(x/E[W])$. ■

Remark 4.13 A sufficient condition for Equation 4.8 is that

$$\sum_{j \neq i} X_j | X_i = 1 \leq_{st} \sum_{j \neq i} X_j, \quad \text{for all } i.$$

As an example where the preceding holds, let $N_1, \ldots, N_r$ have a multinomial distribution, and for given constants $n_1, \ldots, n_r$, let $X_i = I\{N_i \geq n_i\}, i = 1, \ldots, r$.

4.5   Second Moment and Conditional Expectation Inequalities

The second moment inequality gives a lower bound on the probability that a nonnegative random variable is positive.

Proposition 4.14   *Second moment inequality.   For a nonnegative random variable X,*

$$P(X > 0) \geq \frac{(E[X])^2}{E[X^2]}.$$

Proof Using Jensen's inequality in the second line here, we have

$$
\begin{aligned}
E[X^2] &= E[X^2|X>0]P(X>0) \\
&\geq (E[X|X>0])^2 P(X>0) \\
&= \frac{(E[X])^2}{P(X>0)}.
\end{aligned}
$$

∎

When W is the sum of Bernoulli random variables, we can improve the bound of the second moment inequality. So, suppose for the remainder of this section that

$$
W = \sum_{i=1}^{n} X_i,
$$

where X_i is Bernoulli with $E[X_i] = \lambda_i$, $i = 1, \ldots, n$, and $\lambda = \sum_{i=1}^{n} \lambda_i$.

Proposition 4.15 *Conditional expectation inequality.*

$$
P(W > 0) \geq \sum_{i=1}^{n} \frac{\lambda_i}{E[W|X_i = 1]}.
$$

Proof Let $f(0) = 0, f(x) = \frac{1}{x}, x \neq 0$. Lemma 4.11 now gives

$$
\begin{aligned}
P(W > 0) &= \lambda E\left[\frac{1}{W}|X_I = 1\right] \\
&= \lambda \sum_i E\left[\frac{1}{W}|X_I = 1, I = i\right] P(I = i|X_I = 1) \\
&= \sum_i E\left[\frac{1}{W}|X_i = 1\right] \lambda_i \\
&\geq \sum_i \frac{\lambda_i}{E[W|X_i = 1]},
\end{aligned}
$$

where final inequality follows from Jensen's inequality. ∎

Example 4.16 Consider a system consisting of m components, each of which either works or not. Suppose, further, that for given subsets of components $S_j, j = 1, \ldots, n$, none of which is a subset of another, the system functions if all of the components of at least one of these subsets work. If component j independently works with probability α_j, derive a lower bound on the probability the system functions.

Solution   Let X_i equal one if all the components in S_i work, and let it equal zero otherwise, $i = 1, \ldots, n$. Also, let

$$p_i = P(X_i = 1) = \prod_{j \in S_i} \alpha_j.$$

Then, with $W = \sum_{i=1}^{n} X_i$, we have

$$
\begin{aligned}
P(\text{system functions}) &= P(W > 0) \\
&\geq \sum_{i=1}^{n} \frac{p_i}{E[W | X_i = 1]} \\
&= \sum_{i=1}^{n} \frac{p_i}{\sum_{j=1}^{n} P(X_j = 1 | X_i = 1)} \\
&= \sum_{i=1}^{n} \frac{p_i}{1 + \sum_{j \neq i} \prod_{k \in S_j - S_i} \alpha_k},
\end{aligned}
$$

where $S_j - S_i$ consists of all components that are in S_j but not in S_i.   ∎

Example 4.17 Consider a random graph on the set of vertices $\{1, 2, \ldots, n\}$, which is such that each of the $\binom{n}{2}$ pairs of vertices $i \neq j$ is, independently, an edge of the graph with probability p. We are interested in the probability that this graph will be connected, where by connected we mean that for each pair of distinct vertices $i \neq j$ there is a sequence of edges of the form $(i, i_1), (i_1, i_2), \ldots, (i_k, j)$. (That is, a graph is connected if for each each pair of distinct vertices i and j, there is a path from i to j.)

Suppose that

$$p = c \frac{\ln(n)}{n}.$$

We will now show that if $c < 1$, then the probability that the graph is connected goes to zero as $n \to \infty$. To verify this result, consider the number of isolated vertices, where vertex i is said to be isolated if there are no edges of type (i, j). Let X_i be the indicator variable for the event that vertex i is isolated, and let

$$W = \sum_{i=1}^{n} X_i$$

be the number of isolated vertices.
    Now,

$$P(X_i = 1) = (1 - p)^{n-1}.$$

Also,

$$E[W|X_i = 1] = \sum_{j=1}^{n} P(X_j = 1|X_i = 1)$$
$$= 1 + \sum_{j \neq i} (1-p)^{n-2}$$
$$= 1 + (n-1)(1-p)^{n-2}.$$

Because

$$(1-p)^{n-1} = \left(1 - c\frac{\ln(n)}{n}\right)^{n-1}$$
$$\approx e^{-c\ln(n)}$$
$$= n^{-c},$$

the conditional expectation inequality yields that for n large

$$P(W > 0) \geq \frac{n^{1-c}}{1 + (n-1)^{1-c}}.$$

Therefore,

$$c < 1 \Rightarrow P(W > 0\} \to 1 \text{ as } n \to \infty.$$

Because the graph is not connected if $W > 0$, it follows that the graph will almost certainly be disconnected when n is large and $c < 1$. (It can be shown when $c > 1$ that the probability the graph is connected goes to one as $n \to \infty$.) ∎

4.6 Min-Max Identity and Bounds on the Maximum

In this section, we are be interested in obtaining an upper bound on $E[\max_i X_i]$, when $X_1, \ldots, X_n$ are nonnegative random variables. To begin, note that for any nonnegative constant c,

$$\max_i X_i \leq c + \sum_{i=1}^{n} (X_i - c)^+, \tag{4.9}$$

where x^+, the positive part of x, is equal to x if $x > 0$ and is equal to zero otherwise. Taking expectations of the preceding inequality yields

$$E[\max_i X_i] \leq c + \sum_{i=1}^{n} E[(X_i - c)^+].$$

Because $(X_i - c)^+$ is a nonnegative random variable, we have

$$
\begin{aligned}
E[(X_i - c)^+] &= \int_0^\infty P((X_i - c)^+ > x)dx \\
&= \int_0^\infty P(X_i - c > x)dx \\
&= \int_c^\infty P(X_i > y)dy.
\end{aligned}
$$

Therefore,

$$
E[\max_i X_i] \leq c + \sum_{i=1}^n \int_c^\infty P(X_i > y)dy. \tag{4.10}
$$

Because the preceding is true for all $c \geq 0$, the best bound is obtained by choosing the c that minimizes the right side of the preceding. Differentiating, and setting the result equal to zero, shows that the best bound is obtained when c is the value c^* such that

$$
\sum_{i=1}^n P(X_i > c^*) = 1.
$$

Because $\sum_{i=1}^n P(X_i > c)$ is a decreasing function of c, the value of c^* can be easily approximated and then utilized in the inequality in Equation 4.10. It is interesting to note that c^* is such that the expected number of the X_i that exceed c^* is equal to one, which is interesting because the inequality in Equation 4.9 becomes an equality when exactly one of the X_i exceed c.

Example 4.18   Suppose the X_i are independent exponential random variables with rates $\lambda_i, i = 1, \ldots, n$. Then the minimizing value c^* is such that

$$
1 = \sum_{i=1}^n e^{-\lambda_i c^*}
$$

with resulting bound

$$
\begin{aligned}
E[\max_i X_i] &\leq c^* + \sum_{i=1}^n \int_{c^*}^\infty e^{-\lambda_i y}dy \\
&= c^* + \sum_{i=1}^n \frac{1}{\lambda_i} e^{-\lambda_i c^*}.
\end{aligned}
$$

In the special case where the rates are all equal, say $\lambda_i = 1$, then

$$
1 = ne^{-c^*} \quad \text{or} \quad c^* = \ln(n),
$$

and the bound becomes

$$E[\max_i X_i] \leq \ln(n) + 1. \tag{4.11}$$

However, it is easy to compute the expected maximum of a sequence of independent exponentials with a rate of one. Interpreting these random variables as the failure times of n components, we can write

$$\max_i X_i = \sum_{i=1}^n T_i,$$

where T_i is the time between the $(i-1)st$ and the ith failure. Using the lack of memory property of exponentials, it follows that the T_i are independent, with T_i being an exponential random variable with rate $n - i + 1$. (This is because when the $(i - 1)$ failure occurs, the time until the next failure is the minimum of the $n - i + 1$ remaining lifetimes, each of which is exponential with a rate of one.) Therefore, in this case

$$E[\max_i X_i] = \sum_{i=1}^n \frac{1}{n - i + 1} = \sum_{i=1}^n \frac{1}{i}.$$

As it is known that, for n large

$$\sum_{i=1}^n \frac{1}{i} \approx \ln(n) + E,$$

where $E \approx 0.5772$ is *Euler's constant*, we see that the bound yielded by the approach can be accurate. (Also, the bound in Equation 4.11 only requires that the X_i are exponential with a rate of one and not that they are independent.) ∎

The preceding bounds on $E[\max_i X_i]$ only involve the marginal distributions of the X_i. When we have additional knowledge about the joint distributions, we can often do better. To illustrate this, we first need to establish an identity relating the maximum of a set of random variables to the minimums of all the partial sets.

For nonnegative random variables $X_1, \ldots, X_n$, fix x and let A_i denote the event that $X_i > x$. Let

$$W = \max(X_1, \ldots, X_n).$$

Noting that W will be greater than x if and only if at least one of the events A_i occur, we have

$$P(W > x) = P\left(\bigcup_{i=1}^n A_i\right),$$

and the inclusion–exclusion identity gives

$$P(W > x) = \sum_{i=1}^{n} P(A_i) - \sum_{i<j} P(A_i A_j) + \sum_{i<j<k} P(A_i A_j A_k)$$
$$+ \cdots + (-1)^{n+1} P(A_1 \cdots A_n),$$

which can be succinctly written

$$P(W > x) = \sum_{r=1}^{n} (-1)^{r+1} \sum_{i_1 < \cdots < i_r} P(A_{i_1} \cdots A_{i_r}).$$

Now,

$$P(A_i) = P(X_i > x\}$$
$$P(A_i A_j) = P(X_i > x, X_j > x\} = P(\min(X_i, X_j) > x)$$
$$P(A_i A_j A_k) = P(X_i > x, X_j > x, X_k > x\}$$
$$= P(\min(X_i, X_j, X_k) > x),$$

and so on. Thus, we see that

$$P(W > x) = \sum_{r=1}^{n} (-1)^{r+1} \sum_{i_1 < \cdots < i_r} P(\min(X_{i_1}, \ldots, X_{i_r}) > x)).$$

Integrating both sides as x goes from zero to ∞ gives the result:

$$E[W] = \sum_{r=1}^{n} (-1)^{r+1} \sum_{i_1 < \cdots < i_r} E[\min(X_{i_1}, \ldots, X_{i_r})].$$

Moreover, using that going out one term in the inclusion–exclusion identity results in an upper bound on the probability of the union, going out two terms yields a lower bound, going out three terms yields an upper bound, and so on, yields

$$E[W] \leq \sum_{i} E[X_i]$$

$$E[W] \geq \sum_{i} E[X_i] - \sum_{i<j} E[\min(X_i, X_j)]$$

$$E[W] \leq \sum_{i} E[X_i] - \sum_{i<j} E[\min(X_i, X_j)] + \sum_{i<j<k} E[\min(X_i, X_j, X_k)]$$

$$E[W] \geq \ldots,$$

and so on.

Example 4.19 Consider the coupon collectors problem, where each different coupon collected is, independent of past selections, a type i coupon with probability p_i. Suppose we are interested in $E[W]$, where W is the number of coupons we need collect to obtain at least one of each type. Then, letting X_i denote the number that need be collected to obtain a type i coupon, we have that

$$W = \max(X_1, \ldots, X_n),$$

yielding that

$$E[W] = \sum_{r=1}^{n} (-1)^{r+1} \sum_{i_1 < \cdots < i_r} E[\min(X_{i_1}, \ldots, X_{i_r})].$$

Now, $\min(X_{i_1}, \ldots, X_{i_r})$ is the number of coupons that need be collected to obtain any of the types $i_1, \ldots, i_r$. Because each new type collected will be one of these types with probability $\sum_{j=1}^{r} p_{i_j}$, it follows that $\min(X_{i_1}, \ldots, X_{i_r})$ is a geometric random variable with mean $\frac{1}{\sum_{j=1}^{r} p_{i_j}}$. Thus, we obtain the result

$$
\begin{aligned}
E[W] = {} & \sum_{i} \frac{1}{p_i} - \sum_{i<j} \frac{1}{p_i + p_j} \\
& + \sum_{i<j<k} \frac{1}{p_i + p_j + p_k} + \cdots + (-1)^{n+1} \frac{1}{p_1 + \cdots + p_n}.
\end{aligned}
$$

Using the preceding formula for the mean number of coupons needed to obtain a complete set requires summing over 2^n terms, so it is not practical when n is large. Moreover, the bounds obtained by only going out a few steps in the formula for the expected value of a maximum generally turn out to be too loose to be beneficial. However, a useful bound can often be obtained by applying the max-min inequalities to an upper bound for $E[W]$ rather than directly to $E[W]$. We now develop the theory.

For nonnegative random variables $X_1, \ldots, X_n$, let

$$W = \max(X_1, \ldots, X_n).$$

Fix $c \geq 0$, and note the inequality

$$W \leq c + \max((X_1 - c)^+, \ldots, (X_n - c)^+).$$

Now apply the max-min upper bound inequalities to the right side of the preceding, take expectations, and obtain that

$$E[W] \leq c + \sum_{i=1}^{n} E[(X_i - c)^+]$$

$$E[W] \leq c + \sum_{i} E[(X_i - c)^+] - \sum_{i<j} E[\min((X_i - c)^+, (X_j - c)^+)]$$

$$+ \sum_{i<j<k} E[\min((X_i - c)^+, (X_j - c)^+, (X_k - c)^+)],$$

and so on.

Example 4.20   Consider the case of independent exponential random variables $X_1, \ldots, X_n$, all having a rate of one. Then, the preceding gives the bound

$$E[\max_i X_i]$$

$$\leq c + \sum_{i} E[(X_i - c)^+] - \sum_{i<j} E[\min((X_i - c)^+, (X_j - c)^+)]$$

$$+ \sum_{i<j<k} E[\min((X_i - c)^+, (X_j - c)^+, (X_k - c)^+)].$$

To obtain the terms in the three sums of the right-hand side of the preceding, condition, respectively, on whether $X_i > c$, whether $\min(X_i, X_j) > c$, and whether $\min(X_i, X_j, X_k) > c$. This yields

$$E[(X_i - c)^+] = e^{-c}$$

$$E[\min((X_i - c)^+, (X_j - c)^+)] = e^{-2c}\frac{1}{2}$$

$$E[\min((X_i - c)^+, (X_j - c)^+, (X_k - c)^+)] = e^{-3c}\frac{1}{3}.$$

Using the constant $c = \ln(n)$ yields the bound

$$E[\max_i X_i] \leq \ln(n) + 1 - \frac{n(n-1)}{4n^2} + \frac{n(n-1)(n-2)}{18n^3}$$

$$\approx \ln(n) + .806 \qquad \text{for } n \text{ large.}$$

Example 4.21   Let us reconsider Example 4.19, the coupon collector's problem. Let c be an integer. To compute $E[(X_i - c)^+]$, condition on whether a type i coupon is among the first c collected.

$$\begin{aligned}
E[(X_i - c)^+] &= E[(X_i - c)^+ | X_i \leq c]P(X_i \leq c) \\
&\quad + E[(X_i - c)^+ | X_i > c]P(X_i > c) \\
&= E[(X_i - c)^+ | X_i > c](1 - p_i)^c \\
&= \frac{(1 - p_i)^c}{p_i}.
\end{aligned}$$

Similarly,

$$E[\min((X_i - c)^+, (X_j - c)^+)] = \frac{(1 - p_i - p_j)^c}{p_i + p_j}$$

$$E[\min((X_i - c)^+, (X_j - c)^+, (X_k - c)^+)] = \frac{(1 - p_i - p_j - p_k)^c}{p_i + p_j + p_k}.$$

Therefore, for any nonnegative integer c

$$E[W] \le c + \sum_i \frac{(1 - p_i)^c}{p_i}$$

$$E[W] \le c + \sum_i \frac{(1 - p_i)^c}{p_i} - \sum_{i<j} \frac{(1 - p_i - p_j)^c}{p_i + p_j}$$

$$+ \sum_{i<j<k} \frac{(1 - p_i - p_j - p_k)^c}{p_i + p_j + p_k}.$$

4.7 Stochastic Orderings

We say that X is stochastically greater than Y, written $X \ge_{st} Y$, if

$$P(X > t) \ge P(Y > t) \quad \text{for all } t.$$

In this section, we define and compare some other stochastic orderings of random variables.

If X is a nonnegative continuous random variable with distribution function F and density f, then the *hazard rate function* of X is defined by

$$\lambda_X(t) = f(t)/\bar{F}(t),$$

where $\bar{F}(t) = 1 - F(t)$. Interpreting X as the lifetime of an item, then for ϵ small

$$P(t \text{ year old item dies within an additional time } \epsilon)$$
$$= P(X < t + \epsilon | X > t)$$
$$\approx \lambda_X(t)\epsilon.$$

If Y is a nonnegative continuous random variable with distribution function G and density g, say that X is *hazard rate order larger* than Y, written $X \ge_{hr} Y$, if

$$\lambda_X(t) \le \lambda_Y(t) \quad \text{for all } t.$$

Say that X is *likelihood ratio order* larger than Y, written $X \ge_{lr} Y$, if

$$f(x)/g(x) \uparrow x,$$

where f and g are the respective densities of X and Y.

Proposition 4.22

$$X \geq_{lr} Y \Rightarrow X \geq_{hr} Y \Rightarrow X \geq_{st} Y.$$

Proof Let X have density f, and Y have density g. Suppose $X \geq_{lr} Y$. Then, for $y > x$,

$$f(y) = g(y)\frac{f(y)}{g(y)} \geq g(y)\frac{f(x)}{g(x)},$$

implying that

$$\int_x^\infty f(y)dy \geq \frac{f(x)}{g(x)} \int_x^\infty g(y)dy$$

or

$$\lambda_X(x) \leq \lambda_Y(x).$$

To prove the final implication, note first that

$$\int_0^s \lambda_X(t)dt = \int_0^s \frac{f(t)}{\bar{F}(t)} dt = -\log \bar{F}(s)$$

or

$$\bar{F}(s) = e^{-\int_0^s \lambda_X(t)},$$

which immediately shows that $X \geq_{hr} Y \Rightarrow X \geq_{st} Y$. ∎

Define $r_X(t)$, the *reversed hazard rate function* of X, by

$$r_X(t) = f(t)/F(t),$$

and note that

$$\lim_{\epsilon \downarrow 0} \frac{P(t - \epsilon < X | X < t)}{\epsilon} = r_X(t).$$

Say that X is *reverse hazard rate order larger* than Y, written $X \geq_{rh} Y$, if

$$r_X(t) \geq r_Y(t) \quad \text{for all } t.$$

Our next theorem gives some representations of the orderings of X and Y in terms of stochastically larger relations between certain conditional distributions of X and Y. Before presenting it, we introduce the following notation.

Notation For a random variable X and event A, let $[X|A]$ denote a random variable with a distribution that is that of the conditional distribution of X given A.

Theorem 4.23

(a) $X \geq_{hr} Y \Leftrightarrow [X|X > t] \geq_{st} [Y|Y > t]$ *for all* t.

(b) $X \geq_{rh} Y \Leftrightarrow [X|X < t] \geq_{st} [Y|Y < t]$ *for all* t.

(c) $X \geq_{lr} Y \Leftrightarrow [X|s < X < t] \geq_{st} [Y|s < Y < t]$ *for all* $s < t$.

Proof To prove Part (a), let $X_t =_d [X - t|X > t]$ and $Y_t =_d [Y - t|Y > t]$. Noting that

$$\lambda_{X_t}(s) = \begin{cases} 0 & \text{if } s < t \\ \lambda_X(s+t) & \text{if } s \geq t \end{cases}$$

shows that

$$X \geq_{hr} Y \Rightarrow X_t \geq_{hr} Y_t \Rightarrow X_t \geq_{st} Y_t \Rightarrow [X|X > t] \geq_{st} [Y|Y > t].$$

To go the other way, use the identity

$$\bar{F}_{X_t}(\epsilon) = e^{-\int_t^{t+\epsilon} \lambda_X(s)ds}$$

to obtain that

$$[X|X > t] \geq_{st} [Y|Y > t] \Leftrightarrow X_t \geq_{st} Y_t$$
$$\Rightarrow \lambda_X(t) \leq \lambda_Y(t).$$

(b) With $X_t =_d [t - X|X < t]$,

$$\lambda_{X_t}(y) = r_X(t - y) \quad 0 < y < t.$$

Therefore, with $Y_t =_d [t - Y|Y < t]$,

$$X \geq_{rh} Y \Leftrightarrow \lambda_{X_t}(y) \geq \lambda_{Y_t}(y)$$
$$\Rightarrow X_t \leq_{st} Y_t$$
$$\Leftrightarrow [t - X|X < t] \leq_{st} [t - Y|Y < t]$$
$$\Leftrightarrow [X|X < t] \geq_{st} [Y|Y < t].$$

On the other hand,

$$[X|X < t] \geq_{st} [Y|Y < t] \Leftrightarrow X_t \leq_{st} Y_t$$
$$\Rightarrow \int_0^\epsilon \lambda_{X_t}(y)dy \geq \int_0^\epsilon \lambda_{Y_t}(y)dy$$
$$\Leftrightarrow \int_0^\epsilon r_X(t - y)dy \geq \int_0^\epsilon r_Y(t - y)dy$$
$$\Rightarrow r_X(t) \geq r_Y(t).$$

(c)   Let X and Y have respective densities f and g. Suppose $[X|s < X < t] \geq_{st} [Y|s < Y < t]$ for all $s < t$. Letting $s < v < t$, this implies that

$$P(X > v|s < X < t) \geq P(Y > v|s < Y < t)$$

or, equivalently, that

$$\frac{P(v < X < t)}{P(s < X < t)} \geq \frac{P(v < Y < t)}{P(s < Y < t)}$$

or, upon inverting, that

$$1 + \frac{P(s < X \leq v)}{P(v < X < t)} \leq 1 + \frac{P(s < Y \leq v)}{P(v < Y < t)}$$

or, equivalently, that

$$\frac{P(s < X \leq v)}{P(s < Y \leq v)} \leq \frac{P(v < X < t)}{P(v < Y < t)}. \tag{4.12}$$

Letting $v \downarrow s$ in Equation 4.12 yields

$$\frac{f(s)}{g(s)} \leq \frac{P(s < X < t)}{P(s < Y < t)},$$

whereas letting $v \uparrow t$ in Equation 4.12 yields

$$\frac{P(s < X < t)}{P(s < Y < t)} \leq \frac{f(t)}{g(t)}.$$

Thus, $\frac{f(t)}{g(t)} \geq \frac{f(s)}{g(s)}$,   showing that $X \geq_{lr} Y$.

Now suppose that $X \geq_{lr} Y$. Then clearly $[X|s < X < t] \geq_{lr} [Y|s < Y < t]$, implying from Proposition 4.22 that $[X|s < X < t] \geq_{st} [Y|s < Y < t]$. ∎

Corollary 4.24 $X \geq_{lr} Y \Rightarrow X \geq_{rh} Y \Rightarrow X \geq_{st} Y.$

Proof The first implication immediately follows from Parts (b) and (c) of Theorem 4.23. The second implication follows upon taking the limit as $t \to \infty$ in Part (b) of that theorem. ∎

Say that X is an *increasing hazard rate* (IHR) random variable if $\lambda_X(t)$ is nondecreasing in t. (Other terminology is to say that X has an increasing failure rate.)

Proposition 4.25 *Let $X_t =_d [X - t | X > t]$. Then,*
(a) $X_t \downarrow_{st}$ as $t \uparrow$ $\Leftrightarrow$ X is IHR and
(b) $X_t \downarrow_{lr}$ as $t \uparrow$ $\Leftrightarrow$ $\log f(x)$ is concave.

Proof (a) Let $\lambda(y)$ be the hazard rate function of X. Then $\lambda_t(y)$, the hazard rate function of X_t, is given by

$$\lambda_t(y) = \lambda(t + y), \quad y > 0.$$

Hence, if X is IHR then $X_t \downarrow_{hr} t$, implying that $X_t \downarrow_{st} t$. Now, let $s < t$, and suppose that $X_s \geq_{st} X_t$. Then,

$$e^{-\int_t^{t+\epsilon} \lambda(y)dy} = P(X_t > \epsilon) \leq P(X_s > \epsilon) = e^{-\int_s^{s+\epsilon} \lambda(y)dy},$$

showing that $\lambda(t) \geq \lambda(s)$. Thus Part (a) is proved.
(b) Using that the density of X_t is $f_{X_t}(x) = f(x+t)/\bar{F}(t)$ yields, for $s < t$, that

$$
\begin{aligned}
X_s \geq_{lr} X_t \;&\Leftrightarrow\; \frac{f(x+s)}{f(x+t)} \uparrow x \\
&\Leftrightarrow\; \log f(x+s) - \log f(x+t) \uparrow x \\
&\Leftrightarrow\; \frac{f'(x+s)}{f(x+s)} - \frac{f'(x+t)}{f(x+t)} \geq 0 \\
&\Leftrightarrow\; \frac{f'(y)}{f(y)} \downarrow y \\
&\Leftrightarrow\; \frac{d}{dy} \log f(y) \downarrow y \\
&\Leftrightarrow\; \log f(y) \text{ is concave.}
\end{aligned}
$$

■

4.8 Exercises

1. For a nonnegative random variable X, show that $(E[X^n])^{1/n}$ is non-decreasing in n.

2. Let X be as standard normal random variable. Use Corollary 4.4, along with the density

$$g(x) = xe^{-x^2/2}, \quad x > 0$$

to show, for $c > 0$, that
(a) $P(X > c) = \frac{1}{\sqrt{2\pi}} e^{-c^2/2} E_g[\frac{1}{X} | X > c].$

(b)   Show, for any positive random variable W, that

$$E\left[\frac{1}{W}|W > c\right] \le E\left[\frac{1}{W}\right].$$

(c)   Show that

$$P(X > c) \le \frac{1}{2}e^{-c^2/2}.$$

(d)   Show that

$$E_g[X|X > c] = c + e^{c^2/2}\sqrt{2\pi}P(X > c).$$

(e) Use Jensen's inequality, along with the preceding, to show that

$$cP(X > c) + e^{c^2/2}\sqrt{2\pi}(P(X > c))^2 \ge \frac{1}{\sqrt{2\pi}}e^{-c^2/2}.$$

(f)   Argue from Part (e) that $P(X > c)$ must be at least as large as the positive root of the equation

$$cx + e^{c^2/2}\sqrt{2\pi}x^2 = \frac{1}{\sqrt{2\pi}}e^{-c^2/2}.$$

(g)   Conclude that

$$P(X > c) \ge \frac{1}{2\sqrt{2\pi}}\left(\sqrt{c^2 + 4} - c\right)e^{-c^2/2}.$$

3. Let X be a Poisson random variable with mean λ. Show that, for $n \ge \lambda$, the Chernoff bound yields that

$$P(X \ge n) \le \frac{e^{-\lambda}(\lambda e)^n}{n^n}.$$

4. Let $m(t) = E[X^t]$. The *moment bound* states that for $c > 0$

$$P(X \ge c) \le m(t)c^{-t}$$

for all $t > 0$. Show that this result can be obtained from the importance sampling identity.

5. Fill in the details of the proof that, for independent Bernoulli random variables $X_1, \ldots, X_n$, and $c > 0$,

$$P(S - E[S] \le -c) \le e^{-2c^2/n},$$

where $S = \sum_{i=1}^n X_i$.

6. If X is a binomial random variable with parameters n and p, show
   (a) $P(|X - np| \geq c) \leq 2e^{-2c^2/n}$ and
   (b) $P(X - np \geq \alpha np) \leq \exp\{-2np^2\alpha^2\}$.

7. Give the details of the proof of Equation 4.7.

8. Prove that
$$E[f(X)] \geq E[f(E[X|Y])] \geq f(E[X]).$$
   Suppose you want a lower bound on $E[f(X)]$ for a convex function f. The preceding shows that first conditioning on Y and then applying Jensen's inequality to the individual terms $E[f(X)|Y = y]$ results in a larger lower bound than does an immediate application of Jensen's inequality.

9. Let X_i be binary random variables with parameters $p_i, i = 1, \ldots, n$. Let $X = \sum_{i=1}^{n} X_i$, and also let I, independent of the variables $X_1, \ldots, X_n$, be equally likely to be any of the values $1, \ldots, n$. For R independent of I, show that
   (a) $P(I = i|X_I = 1) = p_i/E[X]$,
   (b) $E[XR] = E[X]E[R|X_I = 1]$, and
   (c) $P(X > 0) = E[X] E[\frac{1}{X}|X_I = 1]$.

10. For X_i and X as in Exercise 9, show that
$$\sum_i \frac{p_i}{E[X|X_i = 1]} \geq \frac{(E[X])^2}{E[X^2]}.$$
   Thus, for sums of binary variables, the conditional expectation inequality yields a stronger lower bound than does the second moment inequality.
   Hint: Make use of the results of Exercises 8 and 9.

11. Let X_i be exponential with mean $8 + 2i$, for $i = 1, 2, 3$. Obtain an upper bound on $E[\max X_i]$, and compare it with the exact result when the X_i are independent.

12. Let $U_i, i = 1, \ldots, n$ be uniform $(0, 1)$ random variables. Obtain an upper bound on $E[\max U_i]$, and compare it with the exact result when the U_i are independent.

13. Let U_1 and U_2 be uniform $(0, 1)$ random variables. Obtain an upper bound on $E[\max(U_1, U_2)]$, and show this maximum is obtained when $U_1 = 1 - U_2$.

14. Show that $X \geq_{hr} Y$ if and only if
$$\frac{P(X > t)}{P(X > s)} \geq \frac{P(Y > t)}{P(Y > s)}$$
   for all $s < t$.

15. Let $h(x, y)$ be a real valued function satisfying

$$h(x, y) \geq h(y, x) \quad \text{whenever} \quad x \geq y.$$

(a) Show that if X and Y are independent and $X \geq_{lr} Y$, then $h(X, Y) \geq_{st} h(Y, X)$.
(b) Show by a counterexample that the preceding is not valid under the weaker condition $X \geq_{st} Y$.

16. There are n jobs, with job i requiring a random time X_i to process. The jobs must be processed sequentially. Give a sufficient condition, the weaker the better, under which the policy of processing jobs in the order $1, 2, \ldots, n$ maximizes the probability that at least k jobs are processed by time t for all k and t.

17. Verify Remark 4.13.

5

Markov Chains

5.1 Introduction

This chapter introduces a natural generalization of a sequence of independent random variables, called a Markov chain, where a variable may depend on the immediately preceding variable in the sequence. Named after the 19th century Russian mathematician Andrei Andreyevich Markov, these chains are widely used as simple models of more complex real-world phenomena.

Given a sequence of discrete random variables $X_0, X_1, X_2, \ldots$ taking values in some finite or countably infinite set S, we say that X_n is a *Markov chain* with respect to a filtration $\mathcal{F}_n$ if $X_n \in \mathcal{F}_n$ for all n and, for all $B \subseteq S$, we have the *Markov property*

$$P(X_{n+1} \in B \mid \mathcal{F}_n) = P(X_{n+1} \in B \mid X_n).$$

If we interpret X_n as the state of the chain at time n, then the preceding means that if you know the current state, nothing else from the past is relevant to the future of the Markov chain. That is, given the present state, the future states and the past states are independent. When we let $\mathcal{F}_n = \sigma(X_0, X_1, \ldots, X_n)$, this definition reduces to

$$P(X_{n+1} = j \mid X_n = i, X_{n-1} = i_{n-1}, \ldots, X_0 = i_0)$$
$$= P(X_{n+1} = j \mid X_n = i).$$

If $P(X_{n+1} = j \mid X_n = i)$ is the same for all n, we say that the Markov chain has *stationary transition probabilities*, and we set

$$P_{ij} = P(X_{n+1} = j \mid X_n = i).$$

In this case, the quantities P_{ij} are called the *transition probabilities*, and specifying them along with a probability distribution for the starting state

X_0 is enough to determine all probabilities concerning $X_0, \ldots, X_n$. We will assume from here on that all Markov chains considered have stationary transition probabilities. In addition, unless otherwise noted, we will assume that S, the set of all possible states of the Markov chain, is the set of nonnegative integers.

Example 5.1 *Reflected random walk.* Suppose Y_i are iid Bernoulli(p) random variables, and let $X_0 = 0$ and $X_n = (X_{n-1} + 2Y_n - 1)^+$ for $n = 1, 2, \ldots$. The process X_n, called a *reflected random walk*, can be viewed as the position of a particle at time n such that at each time the particle has a p probability of moving one step to the right and a $1 - p$ probability of moving one step to the left; it is returned to position zero if it ever attempts to move to the left of zero. It is immediate from its definition that X_n is a Markov chain.

Example 5.2 *A non-Markov chain.* Again let Y_i be iid Bernoulli(p) random variables, let $X_0 = 0$, and this time let $X_n = Y_n + Y_{n-1}$ for $n = 1, 2, \ldots$. It's easy to see that X_n is not a Markov chain because $P(X_{n+1} = 2 | X_n = 1, X_{n-1} = 2) = 0$, whereas on the other hand $P(X_{n+1} = 2 | X_n = 1, X_{n-1} = 0) = p$.

5.2   Transition Matrix

The transition probabilities

$$P_{ij} = P(X_1 = j | X_0 = i)$$

are also called the one-step transition probabilities. We define the n-step transition probabilities by

$$P_{ij}^{(n)} = P(X_n = j | X_0 = i).$$

In addition, we define the transition probability matrix

$$\mathbf{P} = \begin{bmatrix} P_{00} & P_{01} & P_{02} & \cdots \\ P_{10} & P_{11} & P_{12} & \cdots \\ \vdots & \vdots & \vdots & \end{bmatrix},$$

and the n-step transition probability matrix

$$\mathbf{P}^{(n)} = \begin{bmatrix} P_{00}^{(n)} & P_{01}^{(n)} & P_{02}^{(n)} & \cdots \\ P_{10}^{(n)} & P_{11}^{(n)} & P_{12}^{(n)} & \cdots \\ \vdots & \vdots & \vdots & \end{bmatrix}.$$

An interesting relation between these matrices is obtained by noting that

$$P_{ij}^{(n+m)} = \sum_k P(X_{n+m} = j | X_0 = i, X_n = k)P(X_n = k | X_0 = i)$$
$$= \sum_k P_{kj}^{(m)} P_{ik}^{(n)}.$$

The preceding are called the *Chapman–Kolmogorov equations*.

If follows from the Chapman–Kolmogorov equations that

$$\mathbf{P}^{(n+m)} = \mathbf{P^n} \times \mathbf{P^m},$$

where $\times$ represents matrix multiplication. Hence,

$$\mathbf{P}^{(2)} = \mathbf{P} \times \mathbf{P},$$

and by induction,

$$\mathbf{P}^{(n)} = \mathbf{P}^n,$$

where the right-hand side represents multiplying the matrix $\mathbf{P}$ by itself n times.

Example 5.3 *Reflected random walk.* A particle starts at position zero and at each time moves one position to the right with probability p and, if the particle is not in position zero, moves one position to the left (or remains in state zero) with probability $1 - p$. The position X_n of the particle at time n forms a Markov chain with transition matrix

$$\mathbf{P} = \begin{bmatrix} 1-p & p & 0 & 0 & 0 & \cdots \\ 1-p & 0 & p & 0 & 0 & \cdots \\ 0 & 1-p & 0 & p & 0 & \cdots \\ \vdots & \vdots & \vdots & & & \end{bmatrix}.$$

Example 5.4 *Two-state Markov chain.* Consider a Markov chain with states zero and one having transition probability matrix

$$\mathbf{P} = \begin{bmatrix} \alpha & 1-\alpha \\ \beta & 1-\beta \end{bmatrix}.$$

The two-step transition probability matrix is given by

$$\mathbf{P}^{(2)} = \begin{bmatrix} \alpha^2 + (1-\alpha)\beta & 1 - \alpha^2 - (1-\alpha)\beta \\ \alpha\beta + \beta(1-\beta) & 1 - \alpha\beta - \beta(1-\beta) \end{bmatrix}.$$

5.3   Strong Markov Property

Consider a Markov chain X_n having one-step transition probabilities P_{ij}, which means that if the Markov chain is in state i at a fixed time n, then the next state will be j with probability P_{ij}. However, it is not necessarily true that if the Markov chain is in state i at a randomly distributed time T, the next state will be j with probability P_{ij}. That is, if T is an arbitrary nonnegative integer valued random variable, it is not necessarily true that $P(X_{T+1} = j | X_T = i) = P_{ij}$. For a simple counterexample, suppose

$$T = \min(n : X_n = i, X_{n+1} = j).$$

Then, clearly,

$$P(X_{T+1} = j | X_T = i) = 1.$$

The idea behind this counterexample is that a general random variable T may depend not only on the states of the Markov chain up to time T but also on future states after time T. Recalling that T is a stopping time for a filtration $\mathcal{F}_n$ if $\{T = n\} \in \mathcal{F}_n$ for every n, we see that for a stopping time the value of T can only depend on the states up to time t. We now show that $P(X_{T+n} = j | X_T = i)$ will equal $P_{ij}^{(n)}$ provided that T is a stopping time.

This is usually called the *strong Markov property* and essentially means a Markov chain "starts over" at stopping times. From here on, we define $\mathcal{F}_T \equiv \{A : A \cap \{T = t\} \in \mathcal{F}_t \text{ for all } t\}$, which intuitively represents any information you would know by time T.

Proposition 5.5 *Strong Markov property. Let $X_n, n \geq 0$, be a Markov chain with respect to the filtration $\mathcal{F}_n$. If $T < \infty$ a.s. is a stopping time with respect to $\mathcal{F}_n$, then*

$$P(X_{T+n} = j | X_T = i, \mathcal{F}_T) = P_{ij}^{(n)}.$$

Proof

$$
\begin{aligned}
P(X_{T+n} = j | X_T = i, \mathcal{F}_T, T = t) &= P(X_{t+n} = j | X_t = i, \mathcal{F}_t, T = t) \\
&= P(X_{t+n} = j | X_t = i, \mathcal{F}_t) \\
&= P_{ij}^{(n)},
\end{aligned}
$$

where the next to last equality used the fact that T is a stopping time to give $\{T = t\} \in \mathcal{F}_t$. ∎

Example 5.6 *Losses in queuing busy periods.* Consider a queuing system where X_n is the number of customers in the system at time n. At each time $n = 1, 2, \ldots$, either a new customer arrives or, if there are any customers present, one departs, with the former happening with probability p. Starting with $X_0 = 1$, let $T = \min\{t > 0 : X_t = 0\}$ be the length of a busy period. Suppose also there is only space for at most m customers in the system. Whenever a customer arrives to find m customers already in the system, the customer is lost and departs immediately. Letting N_m be the number of customers lost during a busy period, compute $E[N_m]$.

Solution Let A be the event that the first arrival occurs before the first departure. We will obtain $E[N_m]$ by conditioning on whether A occurs. Now, when A happens, for the busy period to end we must first wait an interval of time until the system goes back to having a single customer, and then after that we must wait another interval of time until the system becomes completely empty. By the Markov property, the number of losses during the first time interval has distribution N_{m-1} because we are now starting with two customers and therefore with only $m - 1$ spaces for additional customers. The strong Markov property tells us that the number of losses in the second time interval has distribution N_m. We therefore have

$$E[N_m|A] = \begin{cases} E[N_{m-1}] + E[N_m] & \text{if } m > 1 \\ 1 + E[N_m] & \text{if } m = 1, \end{cases}$$

and using $P(A) = p$ and $E[N_m|A^c] = 0$, we have

$$E[N_m] = E[N_m|A]P(A) + E[N_m|A^c]P(A^c)$$
$$= pE[N_{m-1}] + pE[N_m]$$

for $m > 1$ along with

$$E[N_1] = p + pE[N_1]$$

and thus

$$E[N_m] = \left(\frac{p}{1-p}\right)^m.$$

It's interesting to notice that $E[N_m]$ increases in m when $p > 1/2$, decreases when $p < 1/2$, and stays constant for all m when $p = 1/2$. The intuition for the case $p = 1/2$ is that when m increases, losses become less frequent but the busy period becomes longer. ∎

We next apply the strong Markov property to obtain a result for the *cover time*, the time when all states of a Markov chain have been visited.

Proposition 5.7 *Cover times. Given an N-state Markov chain X_n, let $T_i = \min\{n \geq 0 : X_n = i\}$ and let $C = \max_i T_i$ be the cover time. Then $E[C] \leq \sum_{m=1}^{N} \frac{1}{m} \max_{i,j} E[T_j|X_0 = i]$.*

Proof Let $I_1, I_2, ..., I_N$ be a random permutation of the integers $1, 2, ..., N$ chosen so that all possible orderings are equally likely. Letting $T_{I_0} = 0$, and noting that $\max_{j \leq m} T_{I_j} - \max_{j \leq m-1} T_{I_j}$ is the additional time after all states $I_1, ..., I_{m-1}$ have been visited until all states $I_1, ..., I_m$ have been visited, we see that

$$C = \sum_{m=1}^{N} \left(\max_{j \leq m} T_{I_j} - \max_{j \leq m-1} T_{I_j} \right).$$

Thus, we have

$$E[C] = \sum_{m=1}^{N} E \left[\max_{j \leq m} T_{I_j} - \max_{j \leq m-1} T_{I_j} \right]$$

$$= \sum_{m=1}^{N} \frac{1}{m} E \left[\max_{j \leq m} T_{I_j} - \max_{j \leq m-1} T_{I_j} | T_{I_m} > \max_{j \leq m-1} T_{I_j} \right]$$

$$\leq \sum_{m=1}^{N} \frac{1}{m} \max_{i,j} E[T_j | X_0 = i],$$

where the second line follows because all orderings are equally likely and thus $P(T_{I_m} > \max_{j \leq m-1} T_{I_j}) = 1/m$, and the third follows the strong Markov property. ∎

5.4 Classification of States

We say that states i, j of a Markov chain *communicate* with each other, or are in the same *class*, if there are integers n and m such that both $P_{ij}^{(n)} > 0$ and $P_{ji}^{(m)} > 0$ hold. This means that it is possible for the chain to get from i to j and vice versa. A Markov chain is called *irreducible* if all states are in the same class.

For a Markov chain X_n, let

$$T_i = \min\{n > 0 : X_n = i\}$$

be the time until the Markov chain first makes a transition into state i. Using the notation $E_i[\cdots]$ and $P_i(\cdots)$ to denote that the Markov chain starts from state i, let

$$f_i = P_i(T_i < \infty)$$

be the probability that the chain ever makes a transition into state i given that it starts in state i. We say that state i is *transient* if $f_i < 1$ and *recurrent* if $f_i = 1$. Let

$$N_i = \sum_{n=1}^{\infty} I_{\{X_n = i\}}$$

be the total number of transitions into state i. The strong Markov property tells us that, starting in state i,

$$N_i + 1 \sim \text{geometric}(1 - f_i)$$

because each time the chain makes a transition into state i there is, independent of all else, a $(1 - f_i)$ chance it will never return. Consequently,

$$E_i[N_i] = \sum_{n=1}^{\infty} P_{ii}^{(n)}$$

is either infinite or finite depending on whether or not state i is recurrent or transient.

Proposition 5.8 *If state i is recurrent and i communicates with j, then j is also recurrent.*

Proof Because i and j communicate, there exist values n and m such that $P_{ij}^{(n)} P_{ji}^{(m)} > 0$. But for any $k > 0$,

$$P_{jj}^{(n+m+k)} \geq P_{ji}^{(m)} P_{ii}^{(k)} P_{ij}^{(n)},$$

where the preceding follows because $P_{jj}^{(n+m+k)}$ is the probability starting in state j that the chain will be back in j after $n + m + k$ transitions, whereas $P_{ji}^{(m)} P_{ii}^{(k)} P_{ij}^{(n)}$ is the probability of the same event occurring but with the additional condition that the chain must also be in i after the first m transitions and then back in i after an additional k transitions. Summing over k shows that

$$E_j[N_j] \geq \sum_{k} P_{jj}^{(n+m+k)} \geq P_{ji}^{(m)} P_{ij}^{(n)} \sum_{k} P_{ii}^{(k)} = \infty.$$

Thus, j is also recurrent. ∎

Proposition 5.9 *If j is transient, then $\sum_{n=1}^{\infty} P_{ij}^{(n)} < \infty$.*

Proof Note that

$$E_i[N_j] = E_i\left[\sum_{n=0}^{\infty} I_{\{X_n = j\}}\right] = \sum_{n=1}^{\infty} P_{ij}^{(n)}.$$

Let f_{ij} denote the probability that the chain ever makes a transition into j given that it starts at i. Then, conditioning on whether such a transition ever occurs yields, upon using the strong Markov property,

$$E_i[N_j] = (1 + E_j[N_j])f_{ij} < \infty$$

because j is transient. ∎

If i is recurrent, let

$$\mu_i = E_i[T_i]$$

denote the mean number of transitions it takes the chain to return to state i, given it starts in i. We say that a recurrent state i is *null* if $\mu_i = \infty$ and *positive* if $\mu_i < \infty$. In the next section, we will show that positive recurrence is a class property, meaning that if i is positive recurrent and communicates with j then j is also positive recurrence. (This also implies, using that recurrence is a class property, that so is null recurrence.)

5.5   Stationary and Limiting Distributions

For a Markov chain X_n starting in some given state i, we define the *limiting probability* of being in state j to be

$$P_j = \lim_{n \to \infty} P_{ij}^{(n)}$$

if the limit exists and is the same for all i.

It is easy to see that not all Markov chains will have limiting probabilities. For instance, consider the two state Markov chain with $P_{01} = P_{10} = 1$. For this chain, $P_{00}^{(n)}$ will equal one when n is even and zero when n is odd, so it has no limit.

Definition 5.10 *State i of a Markov chain X_n is said to have period d if d is the largest integer having the property that $P_{ii}^{(n)} = 0$ when n is not a multiple of d.*

Proposition 5.11 *If states i and j communicate, then they have the same period.*

Proof Let d_k be the period of state k. Let n, m be such that $P_{ij}^{(n)} P_{ji}^{(m)} > 0$. Now, if $P_{ii}^{(r)} > 0$, then

$$P_{jj}^{(r+n+m)} \geq P_{ji}^{(m)} P_{ii}^{(r)} P_{ij}^{(n)} > 0.$$

So d_j divides $r + n + m$. Moreover, because

$$P_{ii}^{(2r)} \geq P_{ii}^{(r)} P_{ii}^{(r)} > 0,$$

the same argument shows that d_j also divides $2r + n + m$; therefore d_j divides $2r + n + m - (r + n + m) = r$. Because d_j divides r whenever $P_{ii}^{(r)} > 0$, it follows that d_j divides d_i. But the same argument can now be used to show that d_i divides d_j. Hence, $d_i = d_j$. ∎

It follows from the preceding that all states of an irreducible Markov chain have the same period. If the period is one, we say that the chain is *aperiodic*. It's easy to see that only aperiodic chains can have limiting probabilities.

Intimately linked to limiting probabilities are stationary probabilities. The probability vector $\pi_i, i \in S$ is said to be a *stationary probability vector* for the Markov chain if

$$\pi_j = \sum_i \pi_i P_{ij} \quad \text{for all} \quad j$$

$$\sum_j \pi_j = 1.$$

Its name arises from the fact that if the X_0 is distributed according to a stationary probability vector $\{\pi_i\}$ then

$$P(X_1 = j) = \sum_i P(X_1 = j | X_0 = i)\pi_i = \sum_i \pi_i P_{ij} = \pi_j$$

and, by a simple induction argument,

$$P(X_n = j) = \sum_i P(X_n = j | X_{n-1} = i)P(X_{n-1} = i) = \sum_i \pi_i P_{ij} = \pi_j.$$

Consequently, if we start the chain with a stationary probability vector then $X_n, X_{n+1}, \dots$ has the same probability distribution for all n.

The following result will be needed later.

Proposition 5.12 *An irreducible transient Markov chain does not have a stationary probability vector.*

Proof Assume there is a stationary probability vector $\pi_i, i \geq 0$, and take it to be the probability mass function of X_0. Then, for any j

$$\pi_j = P(X_n = j) = \sum_i \pi_i P_{ij}^{(n)}.$$

Consequently, for any m

$$\pi_j = \lim_{n\to\infty} \sum_i \pi_i P_{ij}^{(n)}$$

$$\leq \lim_{n\to\infty} \left(\sum_{i\leq m} \pi_i P_{ij}^{(n)} + \sum_{i>m} \pi_i \right)$$

$$= \sum_{i\leq m} \pi_i \lim_{n\to\infty} P_{ij}^{(n)} + \sum_{i>m} \pi_i$$

$$= \sum_{i>m} \pi_i,$$

where the final equality used that $\sum_j P_{ij}^{(n)} < \infty$ because j is transient, implying that $\lim_{n\to\infty} P_{ij}^{(n)} = 0$. Letting $m \to \infty$ shows that $\pi_j = 0$ for all j, contradicting the fact that $\sum_j \pi_j = 1$. Thus, assuming a stationary probability vector results in a contradiction, proving the result. ∎

The following theorem is of key importance.

Theorem 5.13 *An irreducible Markov chain has a stationary probability vector* $\{\pi_i\}$ *if and only if all states are positive recurrent. The stationary probability vector is unique and satisfies*

$$\pi_j = 1/\mu_j.$$

Moreover, if the chain is aperiodic then

$$\pi_j = \lim_n P_{ij}^{(n)}.$$

To prove the preceding theorem, we will make use of a couple of lemmas.

Lemma 5.14 *For an irreducible Markov chain, if there exists a stationary probability vector* $\{\pi_i\}$, *then all states are positive recurrent. Moreover, the stationary probability vector is unique and satisfies*

$$\pi_j = 1/\mu_j.$$

Proof Let π_j be stationary probabilities, and suppose that $P(X_0 = j) = \pi_j$ for all j. We first show that $\pi_i > 0$ for all i. To verify this, suppose that $\pi_k = 0$. Now for any state j, because the chain is irreducible, there is an n such that $P_{jk}^{(n)} > 0$. Because X_0 is determined by the stationary probabilities,

$$\pi_k = P(X_n = k) = \sum_i \pi_i P_{ik}^{(n)} \geq \pi_j P_{jk}^{(n)}.$$

Consequently, if $\pi_k = 0$ then so is π_j. Because j was arbitrary, that means that if $\pi_i = 0$ for any i, then $\pi_i = 0$ for all i. But that would contradict the fact that $\sum_i \pi_i = 1$. Hence, any stationary probability vector for an irreducible Markov chain must have all positive elements.

Now, recall that $T_j = \min(n > 0 : X_n = j)$. So,

$$
\begin{aligned}
\mu_j &= E[T_j | X_0 = j] \\
&= \sum_{n=1}^{\infty} P(T_j \geq n | X_0 = j) \\
&= \sum_{n=1}^{\infty} \frac{P(T_j \geq n, X_0 = j)}{P(X_0 = j)}.
\end{aligned}
$$

Because X_0 is chosen according to the stationary probability vector $\{\pi_i\}$, this gives

$$
\pi_j \mu_j = \sum_{n=1}^{\infty} P(T_j \geq n, X_0 = j). \tag{5.1}
$$

Now,

$$
P(T_j \geq 1, X_0 = j) = P(X_0 = j) = \pi_j,
$$

and for $n \geq 2$,

$$
\begin{aligned}
P(T_j &\geq n, X_0 = j) \\
&= P(X_i \neq j, 1 \leq i \leq n-1, X_0 = j) \\
&= P(X_i \neq j, 1 \leq i \leq n-1) - P(X_i \neq j, 0 \leq i \leq n-1) \\
&= P(X_i \neq j, 1 \leq i \leq n-1) - P(X_i \neq j, 1 \leq i \leq n),
\end{aligned}
$$

where the final equality used that $X_0, \ldots, X_{n-1}$ has the same probability distribution as $X_1, \ldots, X_n$, when X_0 is chosen according to the stationary probabilities. Substituting these results into Equation 5.1 yields

$$
\pi_j \mu_j = \pi_j + P(X_1 \neq j) - \lim_n P(X_i \neq j, 1 \leq i \leq n).
$$

But the existence of a stationary probability vector implies that the Markov chain is recurrent and that $\lim_n P(X_i \neq j, 1 \leq i \leq n) = P(X_i \neq j, \text{for all } i \geq 1) = 0$. Because $P(X_1 \neq j) = 1 - \pi_j$, we thus obtain

$$
\pi_j = 1/\mu_j,
$$

showing that there is at most one stationary probability vector. In addition, because all $\pi_j > 0$, we have that all $\mu_j < \infty$, showing that all states of the chain are positive recurrent. ■

Lemma 5.15 *If some state of an irreducible Markov chain is positive recurrent, then there exists a stationary probability vector.*

Proof Suppose state k is positive recurrent. Thus,

$$\mu_k = E_k[T_k] < \infty.$$

Say that a new cycle begins every time the chain makes a transition into state k. For any state j, let A_j denote the amount of time the chain spends in state j during a cycle. Then

$$
\begin{aligned}
E[A_j] &= E_k\left[\sum_{n=0}^{\infty} I_{\{X_n=j,T_k>n\}}\right] \\
&= \sum_{n=0}^{\infty} E_k[I_{\{X_n=j,T_k>n\}}] \\
&= \sum_{n=0}^{\infty} P_k(X_n = j, T_k > n).
\end{aligned}
$$

We claim that $\pi_j \equiv E[A_j]/\mu_k$, $j \geq 0$, is a stationary probability vector. Because $E[\sum_j A_j]$ is the expected time of a cycle, it must equal $E_k[T_k]$, showing that

$$\sum_j \pi_j = 1.$$

Moreover, for $j \neq k$

$$
\begin{aligned}
\mu_k \pi_j &= \sum_{n \geq 0} P_k(X_n = j, T_k > n) \\
&= \sum_{n \geq 1} \sum_i P_k(X_n = j, T_k > n-1, X_{n-1} = i) \\
&= \sum_{n \geq 1} \sum_i P_k(T_k > n-1, X_{n-1} = i) \\
&\qquad \times P_k(X_n = j | T_k > n-1, X_{n-1} = i) \\
&= \sum_i \sum_{n \geq 1} P_k(T_k > n-1, X_{n-1} = i) P_{ij} \\
&= \sum_i \sum_{n \geq 0} P_k(T_k > n, X_n = i) P_{ij} \\
&= \sum_i E[A_i] P_{ij} \\
&= \mu_k \sum_i \pi_i P_{ij}.
\end{aligned}
$$

Finally,

$$\sum_i \pi_i P_{ik} = \sum_i \pi_i \left(1 - \sum_{j \neq k} P_{ij} \right)$$
$$= 1 - \sum_{j \neq k} \sum_i \pi_i P_{ij}$$
$$= 1 - \sum_{j \neq k} \pi_j$$
$$= \pi_k,$$

and the proof is complete. ∎

Note that Lemmas 5.14 and 5.15 imply the following.

Corollary 5.16 *If one state of an irreducible Markov chain is positive recurrent, then all states are positive recurrent.*

We are now ready to prove Theorem 5.13.

Proof All that remains to be proven is that if the chain is aperiodic, as well as irreducible and positive recurrent, then the stationary probabilities are also limiting probabilities. To prove this, let $\pi_i, i \geq 0$, be stationary probabilities. Let $X_n, n \geq 0$ and $Y_n, n \geq 0$ be independent Markov chains, both with transition probabilities $P_{i,j}$, but with $X_0 = i$ and with $P(Y_0 = i) = \pi_i$. Let

$$N = \min(n : X_n = Y_n).$$

We first show that $P(N < \infty) = 1$. To do so, consider the Markov chain with a state at time n that is (X_n, Y_n) and thus has transition probabilities $P_{(i,j),(k,r)} = P_{ik} P_{jr}$.

That this chain is irreducible can be seen by the following argument. Because $\{X_n\}$ is irreducible and aperiodic, it follows that for any state i there are relatively prime integers n, m such that $P_{ii}^{(n)} P_{ii}^{(m)} > 0$. But any sufficiently large integer can be expressed as a linear combination of relatively prime integers, implying that there is an integer N_i such that

$$P_{ii}^{(n)} > 0 \quad \text{for all } n > N_i.$$

Because i and j communicate, this implies the existence of an integer $N_{i,j}$ such that

$$P_{ij}^{(n)} > 0 \quad \text{for all } n > N_{i,j}.$$

Hence,

$$P_{(i,k),(j,r)}^{(n)} = P_{ij}^{(n)} P_{kr}^{(n)} > 0 \quad \text{for all sufficiently large } n,$$

which shows that the vector chain (X_n, Y_n) is irreducible.

In addition, we claim that $\pi_{i,j} = \pi_i \pi_j$ is a stationary probability vector, which is seen from

$$\pi_i \pi_j = \sum_k \pi_k P_{k,i} \sum_r \pi_r P_{r,j} = \sum_{k,r} \pi_k \pi_r P_{k,i} P_{r,j}.$$

By Lemma 5.14, this shows that the vector Markov chain is positive recurrent, so $P(N < \infty) = 1$ and thus $\lim_n P(N > n) = 0$.

Now, let $Z_n = X_n$ if $n < N$ and let $Z_n = Y_n$ if $n \geq N$. It is easy to see that $Z_n, n \geq 0$, is also a Markov chain with transition probabilities $P_{i,j}$ and has $Z_0 = i$. Now

$$
\begin{aligned}
P_{i,j}^{(n)} &= P(Z_n = j) \\
&= P(Z_n = j, N \leq n) + P(Z_n = j, N > n) \\
&= P(Y_n = j, N \leq n) + P(Z_n = j, N > n) \\
&\leq P(Y_n = j) + P(N > n) \\
&= \pi_j + P(N > n).
\end{aligned}
\tag{5.2}
$$

On the other hand,

$$
\begin{aligned}
\pi_j &= P(Y_n = j) \\
&= P(Y_n = j, N \leq n) + P(Y_n = j, N > n) \\
&= P(Z_n = j, N \leq n) + P(Y_n = j, N > n) \\
&\leq P(Z_n = j) + P(N > n) \\
&= P_{ij}^{(n)} + P(N > n).
\end{aligned}
\tag{5.3}
$$

Hence, from Equations 5.2 and 5.3, we see that

$$\lim_n P_{ij}^{(n)} = \pi_j.$$

∎

Remark 5.17 It follows from Theorem 5.13 that if we have an irreducible Markov chain, and we can find a solution of the stationarity equations

$$\pi_j = \sum_i \pi_i P_{ij} \quad j \geq 0$$

$$\sum_i \pi_i = 1,$$

then the Markov chain is positive recurrent, and the π_i are the unique stationary probabilities. If, in addition, the chain is aperiodic, then the π_i are also limiting probabilities.

Remark 5.18 Because μ_i is the mean number of transitions between successive visits to state i, it is intuitive (and will be formally proven in Chapter 6 on renewal theory) that the long run proportion of time that the chain spends in state i is equal to $1/\mu_i$. Hence, the stationary probability π_i is equal to the long-run proportion of time that the chain spends in state i.

Definition 5.19 *A positive recurrent, aperiodic, irreducible Markov chain is called an ergodic Markov chain.*

Definition 5.20 *A positive recurrent irreducible Markov chain with an initial state that is distributed according to its stationary probabilities is called a stationary Markov chain.*

5.6 Time Reversibility

A stationary Markov chain X_n is called *time reversible* if

$$P(X_n = j | X_{n+1} = i) = P(X_{n+1} = j | X_n = i) \quad \text{for all } i, j.$$

By the Markov property, we know that the processes $X_{n+1}, X_{n+2}, \dots$ and $X_{n-1}, X_{n-2}, \dots$ are conditionally independent given X_n, so it follows that the reversed process $X_{n-1}, X_{n-2}, \dots$ will also be a Markov chain having transition probabilities

$$P(X_n = j | X_{n+1} = i) = \frac{P(X_n = j, X_{n+1} = i)}{P(X_{n+1} = i)}$$
$$= \frac{\pi_j P_{ji}}{\pi_i},$$

where π_i and P_{ij} respectively denote the stationary probabilities and the transition probabilities for the Markov chain X_n. Thus, an equivalent definition for X_n being time reversible is if

$$\pi_i P_{ij} = \pi_j P_{ji} \quad \text{for all } i, j.$$

Intuitively, a Markov chain is time reversible if it looks the same running backward as it does running forward. It also means that the rate of transitions from i to j – namely, $\pi_i P_{ij}$ – is the same as the rate of transitions from j to i – namely, $\pi_j P_{ji}$. This happens if there are no "loops" for which a Markov chain is more likely in the long run to go in one direction compared with the other direction. We illustrate this with examples.

Example 5.21 *Random walk on the circle.* Consider a particle that moves around n positions on a circle numbered $1, 2, ..., n$ according to transition probabilities $P_{i,i+1} = p = 1 - P_{i+1,i}$ for $1 \le i < n$ and $P_{n,1} = p = 1 - P_{1,n}$. Let X_n be the position of the particle at time n. Regardless of p, it is easy to see that the stationary probabilities are $\pi_i = 1/n$. Now, for $1 \le i < n$ we have $\pi_i P_{i,i+1} = p/n$ and $\pi_{i+1} P_{i+1,i} = (1-p)/n$ (and also $\pi_n P_{n,1} = p/n$ and $\pi_1 P_{1,n} = (1-p)/n$). If $p = 1/2$ these will all be equal and X_n will be time reversible. On the other hand, if $p \ne 1/2$ these will not be equal and X_n will not be time reversible.

It can be much easier to verify the stationary probabilities for a time reversible Markov chain than for a Markov chain that is not time reversible. Verifying the stationary probabilities π_i for a Markov chain involves checking $\sum_i \pi_i = 1$ and, for all j,

$$\pi_j = \sum_i \pi_i P_{ij}.$$

For a time-reversible Markov chain, it only requires checking that $\sum_i \pi_i = 1$ and

$$\pi_i P_{ij} = \pi_j P_{ji}$$

for all i, j because if the preceding holds, then summing both sides over j yields

$$\pi_i \sum_j P_{ij} = \sum_j \pi_j P_{ji}$$

or

$$\pi_i = \sum_j \pi_j P_{ji}.$$

This can be convenient in some cases, and we illustrate one next.

Example 5.22 *Random walk on a graph.* Consider a particle moving on a graph that consists of nodes and edges, and let d_i be the number of edges emanating from node i. If X_n is the location of the particle at time n, let $P(X_{n+1} = j | X_n = i) = 1/d_i$ if there is an edge connecting node i and node j. This means that, when at a given node, the random walker's next step is equally likely to be to any of the nodes that are connected by an edge. If D is the total number of edges that appear in the graph, we will show that the stationary probabilities are given by $\pi_i = \frac{d_i}{2D}$.

Solution Checking that $\pi_i P_{ij} = \pi_j P_{ji}$ holds for the claimed solution, we see that this requires that

$$\frac{d_i}{2D} \frac{1}{d_i} = \frac{d_j}{2D} \frac{1}{d_j}.$$

It thus follows, because $\sum_i \pi_i = \frac{\sum_i d_i}{2D} = 1$, that the Markov chain is time reversible with the given stationary probabilities. ■

5.7   A Mean Passage Time Bound

Consider now a Markov chain with a state space that is the set of nonnegative integers and is such that

$$P_{ij} = 0, \ 0 \le i < j. \tag{5.4}$$

That is, the state of the Markov chain can never strictly increase. Suppose we are interested in bounding the expected number of transitions it takes such a chain to go from state n to state zero. To obtain such a bound, let D_i be the amount by which the state decreases when a transition from state i occurs so that

$$P(D_i = k) = P_{i,i-k}, \ 0 \le k \le i.$$

The following proposition yields the bound.

Proposition 5.23 *Let N_n denote the number of transitions it takes a Markov chain satisfying Equation 5.4 to go from state n to state zero. If for some nondecreasing function $d_i, i > 0$, we have that $E[D_i] \ge d_i$, then*

$$E[N_n] \le \sum_{i=1}^{n} 1/d_i.$$

Proof The proof is by induction on n. It is true when $n = 1$, because N_1 is geometric with mean

$$E[N_1] = \frac{1}{P_{1,0}} = \frac{1}{E[D_1]} \le \frac{1}{d_1}.$$

So, assume that $E[N_k] \le \sum_{i=1}^{k} 1/d_i$, for all $k < n$. To bound $E[N_n]$, we condition on the transition out of state n and use the induction hypothesis in the first inequality in the following to get

$$E[N_n]$$

$$= \sum_{j=0}^{n} E[N_n | D_n = j] P(D_n = j)$$

$$= 1 + \sum_{j=0}^{n} E[N_{n-j}] P(D_n = j)$$

$$= 1 + P_{n,n} E[N_n] + \sum_{j=1}^{n} E[N_{n-j}] P(D_n = j)$$

$$\leq 1 + P_{n,n} E[N_n] + \sum_{j=1}^{n} P(D_n = j) \sum_{i=1}^{n-j} 1/d_i$$

$$= 1 + P_{n,n} E[N_n] + \sum_{j=1}^{n} P(D_n = j) \left[\sum_{i=1}^{n} 1/d_i - \sum_{i=n-j+1}^{n} 1/d_i \right]$$

$$\leq 1 + P_{n,n} E[N_n] + \sum_{j=1}^{n} P(D_n = j) \left[\sum_{i=1}^{n} 1/d_i - j/d_n \right],$$

where the last line follows because d_i is nondecreasing. Continuing from the previous line, we get

$$= 1 + P_{n,n} E[N_n] + (1 - P_{n,n}) \sum_{i=1}^{n} 1/d_i - \frac{1}{d_n} \sum_{j=1}^{n} j P(D_n = j)$$

$$= 1 + P_{n,n} E[N_n] + (1 - P_{n,n}) \sum_{i=1}^{n} 1/d_i - \frac{E[D_n]}{d_n}$$

$$\leq P_{n,n} E[N_n] + (1 - P_{n,n}) \sum_{i=1}^{n} 1/d_i,$$

which completes the proof. ∎

Example 5.24 At each stage, each of a set of balls is independently put in one of n urns, with each ball being put in urn i with probability p_i, $\sum_{i=1}^{n} p_i = 1$. After this is done, all of the balls in the same urn are coalesced into a single new ball, with this process continually repeated until a single ball remains. Starting with N balls, we would like to bound the mean number of stages needed until a single ball remains.

We can model the preceding as a Markov chain $\{X_k, k \geq 0\}$, with a state that is the number of balls that remain in the beginning of a stage. Because the number of balls that remain after a stage beginning with i balls is equal to the number of nonempty urns when these i balls are distributed, it follows that

$$E[X_{k+1}|X_k = i] = E\left[\sum_{j=1}^{n} I\{\text{urn } j \text{ is nonempty}\}|X_k = i\right]$$

$$= \sum_{j=1}^{n} P(\text{urn } j \text{ is nonempty}|X_k = i)$$

$$= \sum_{j=1}^{n} [1 - (1 - p_j)^i].$$

Hence, $E[D_i]$, the expected decrease from state i is

$$E[D_i] = i - n + \sum_{j=1}^{n} (1 - p_j)^i.$$

Because

$$E[D_{i+1}] - E[D_i] = 1 - \sum_{j=1}^{n} p_j (1 - p_j)^i > 0,$$

it follows from Proposition 5.23 that the mean number of transitions to go from state N to state one satisfies

$$E[N_n] \le \sum_{i=2}^{n} \frac{1}{i - n + \sum_{j=1}^{n} (1 - p_j)^i}.$$

5.8 Gambler's Ruin

Consider a gambler who in each round of a game has a probability p of winning one dollar and a probability $q = 1 - p$ of losing one dollar, with the outcomes of successive rounds being independent. Letting X_n be the net winnings of the gambler after the nth round, the process $\{X_n, n = 0, 1, 2, \ldots\}$ is a Markov chain, also called a random walk, with transition probabilities $p_{i,i+1} = 1 - p_{i,i-1} = p$ for integers i.

Letting $T_p = \min\{t > 0 : X_t = m \text{ or } X_t = -m\}$ be the number of rounds until the gambler's net winnings reaches either $-m$ or $+m$ starting from $X_0 = 0$, we will show that if $1 \ge p_1 \ge p_2 \ge 1/2$ then

$$T_{p_1} \le_{st} T_{p_2}, \tag{5.5}$$

meaning that this duration is stochastically longer when the rounds are more fair.

To verify the preceding result, we first show that $|X_n|, n \geq 0$ is a Markov chain.

To do so, suppose we are given that $|X_0| = x_0, |X_1| = x_1, \ldots, |X_{2n+i}| = x_{2n+i}$, where $x_0 = 0, x_{2n+i} = i$. To determine the conditional probability distribution of $|X_{2n+i+1}|$, we will first determine

$$P(X_{2n+i} = i \mid |X_r| = x_r, r = 0, \ldots, 2n + i).$$

To do so, let

$$j = \max\{k : 0 \leq k \leq 2n + i : x_k = 0\}$$

and note that j is an even integer. Because $X_j = 0$ it follows that

$$P(X_{2n+i} = i \mid |X_r| = x_r, r = 0, \ldots, 2n + i)$$
$$= P(X_{2n+i} = i \mid |X_r| = x_r, r = j, \ldots, 2n + i)$$

Because $x_j = 0, x_r \neq 0, r = j + 1, \ldots, 2n + i$, it follows that there are only two possible values of the sequence $X_{j+1}, \ldots, X_{2n+i}$, with the first occurring if the sequence results from $n - j/2 + i$ up moves and $n - j/2$ down moves, and the second if the reverse occurs. Hence,

$$P(X_{2n+i} = i \mid |X_r| = x_r, r = 0, \ldots, 2n + i)$$
$$= \frac{p^{n-j/2+i}q^{n-j/2}}{p^{n-j/2+i}q^{n-j/2} + q^{n-j/2+i}p^{n-j/2}}$$
$$= \frac{p^i}{p^i + q^i}$$

Conditioning on whether $X_{2n+i} = i$ or $X_{2n+i} = -i$ now gives that

$$P(|X_{2n+i+1}| = i + 1 \mid |X_r| = x_r, r = 0, \ldots, 2n + i) = \frac{p^{i+1} + q^{i+1}}{p^i + q^i}$$

As it is easily shown that the preceding transition probability is increasing in p when $p \geq 1/2$, it follows that if we have two versions of the random walk X_n and X_n' respectively with upward probabilities p_1 and p_2 with $1 \geq p_1 \geq p_2 \geq 1/2$, the upward transition probabilities for the Markov chain $|X_n|$ are always at least as large as for $|X_n'|$. This means we can create a coupling where $|X_n| \geq |X_n'|$ for all n by letting the two Markov chains step independently when at different levels, and when at the same level we can couple their next steps so that the latter never ends up above the former; the two chains will never cross when stepping from different levels because they always are an even number of steps apart. This means the former reaches m sooner and (5.5) holds.

5.9 Exercises

1. Let f_{ij} denote the probability that the Markov chain ever makes a transition into state j given that it starts in state i. Show that if i is recurrent and communicates with j then $f_{ij} = 1$.

2. Show that a recurrent class of states of a Markov chain is a closed class, in the sense that if i is recurrent and i does not communicate with j then $P_{ij} = 0$.

3. The one-dimensional simple random walk is the Markov chain $X_n, n \geq 0$, with states that are all the integers and that has the transition probabilities

$$P_{i,i+1} = 1 - P_{i,i-1} = p.$$

Show that this chain is recurrent when $p = 1/2$ and transient for all $p \neq 1/2$. When $p = 1/2$, the chain is called the one-dimensional simple symmetric random walk.

Hint: Make use of Stirling's approximation, which states that

$$n! \sim n^{n+1/2} e^{-n} \sqrt{2\pi},$$

where we say that $a_n \sim b_n$ if $\lim_{n\to\infty} a_n/b_n = 1$. You can also use the fact that if $a_n > 0, b_n > 0$ for all n, then $a_n \sim b_n$ implies that $\sum_n a_n < \infty$ if and only if $\sum_n b_n < \infty$.

4. The two-dimensional simple symmetric random walk moves on a two-dimensional grid according to the transition probabilities

$$P_{(i,j),(i,j+1)} = P_{(i,j),(i+1,j)} = P_{(i,j),(i-1,j)} = P_{(i,j),(i,j-1)} = 1/4.$$

Show that this Markov chain is recurrent.

5. Define the three-dimensional simple symmetric random walk, and then show that it is transient.

6. Given a finite-state Markov chain X_n, let $T_i = \min\{n \geq 0 : X_n = i\}$ and $C = \max_i T_i$.
   (a) Show that for any subset of states A

$$\min_i E[C|X_0 = i] \geq \sum_{m=1}^{|A|} \frac{1}{m} \min_{i \in A, j \in A} E_i[T_j],$$

where $|A|$ denotes the number of elements in A.
(b) Obtain a lower bound for the mean number of flips required until all 2^k patterns of length k have appeared when a fair coin is repeatedly flipped.

7. Consider a Markov chain with a state space that is the set of nonnegative integers. Suppose its transition probabilities are given by

$$P_{0,i} = p_i, \ i \geq 0, \quad P_{i,i-1} = 1, \ i > 0,$$

where $\sum_i i p_i < \infty$. Find the limiting probabilities for this Markov chain.

8. Consider a Markov chain with states $0, 1, \ldots, N$ and transition probabilities

$$P_{0N} = 1, \quad P_{ij} = 1/i, \, i > 0, \, j < i.$$

That is, from state zero the chain always goes to state N, and from state $i > 0$ it is equally likely to go to any lower numbered state. Find the limiting probabilities of this chain.

9. Consider a Markov chain with states $0, 1, \ldots, N$ and transition probabilities

$$P_{i,i+1} = p = 1 - P_{i,i-1}, \, i = 1, \ldots, N - 1$$

$$P_{0,0} = P_{N,N} = 1.$$

Suppose that $X_0 = i$, where $0 < i < N$. Argue that, with a probability of one, the Markov chain eventually enters either state zero or N. Derive the probability it enters state N before state zero. This is called the *gambler's ruin probability*.

10. If X_n is a stationary ergodic Markov chain, show that $X_1, X_2, \ldots$ is an ergodic sequence.

11. Suppose $X_1, X_2, \ldots$ are iid integer valued random variables with $M_n = \max_{i \le n} X_i$. Is M_n necessarily a Markov chain? If yes, give its transition probabilities; if no, construct a counterexample.

12. Suppose X_n is a finite-state stationary Markov chain, and let $T = \min\{n > 0 : X_n = X_0\}$. Compute $E[T]$.

13. Given an irreducible Markov chain with transition probabilities P_{ij} and any positive probability vector $\{\pi_i\}$ for these states, show that the Markov chain with transition probabilities $Q_{ij} = \min(P_{ij}, \pi_j P_{ji}/\pi_i)$ if $i \ne j$ and $Q_{ii} = 1 - \sum_{j \ne i} Q_{ij}$ is time reversible and has stationary distribution $\{\pi_i\}$.

14. Consider a time-reversible Markov chain with transition probabilities P_{ij} and stationary probabilities π_i. If A is a set of states of this Markov chain, then we define the A-truncated chain as being a Markov chain with a set of states that is A and with transition probabilities P_{ij}^A, $i, j \in A$, that are given by

$$P_{ij}^A = \begin{cases} P_{ij} & \text{if } j \ne i \\ P_{ii} + \sum_{k \notin A} P_{ik} & \text{if } j = i. \end{cases}$$

If this truncated chain is irreducible, show that it is time reversible, with stationary probabilities

$$\pi_i^A = \pi_i \Big/ \sum_{j \in A} \pi_j, \quad i \in A.$$

15. A collection of M balls are distributed among m urns. At each stage, one of the balls is randomly selected, taken from whatever urn it is in and then randomly placed in one of the other $m-1$ urns. Consider the Markov chain with a state that is at any time the vector $(n_1, n_2, \ldots, n_m)$ where n_i is the number of balls in urn i. Show that this Markov chain is time reversible and find its stationary probabilities.

16. Let $\mathbf{Q}$ be an irreducible symmetric transition probability matrix on the states $1, \ldots, n$. That is,

$$Q_{ij} = Q_{ji}, \quad i, j = 1, \ldots, n.$$

Let $b_i, i = 1, \ldots, n$ be specified positive numbers, and consider a Markov chain with transition probabilities

$$P_{ij} = Q_{ij} \frac{b_j}{b_i + b_j}, \; j \neq i$$

$$P_{ii} = 1 - \sum_{j \neq i} P_{ij}.$$

Show that this Markov chain is time reversible with stationary probabilities

$$\pi_i = \frac{b_i}{\sum_{j=1}^n b_j}, \quad i = 1, \ldots, n.$$

17. Consider a Markov chain with a state space that is the set of positive integers and with transition probabilities that are

$$P_{1,1} = 1, \quad P_{ij} = \frac{1}{i-1}, 1 \leq j < i, i > 1.$$

Show that the bound on the mean number of transitions to go from state n to state one given by Proposition 5.23 is approximately twice the actual mean number.

6

Renewal Theory

6.1 Introduction

A counting process with a sequence of interevent times that are iid is called a renewal process. More formally, let $X_1, X_2, \ldots$ be a sequence of iid nonnegative random variables having distribution function F. Assume that $F(0) \neq 1$, so that the X_i are not identically zero, and set

$$S_0 = 0$$
$$S_n = \sum_{i=1}^{n} X_i, \quad n \geq 1.$$

With

$$N(t) = \sup(n : S_n \leq t),$$

the process $\{N(t), t \geq 0\}$ is called a *renewal process*.

If we suppose that events are occurring in time and we interpret X_n as the time between the $(n-1)$ and the nth event, then S_n is the time of the nth event, and $N(t)$ represents the number of events that occur before or at time t. An event is also called a *renewal* because, if we consider the time of occurrence of an event as the new origin, then because the X_i are iid, the process of future events is also a renewal process with interarrival distribution F. Thus, the process probabilistically restarts, or *renews*, whenever an event occurs.

Let $\mu = E[X_i]$. Because $P(X_i \geq 0) = 1$ and $P(X_i = 0) < 1$, it follows that $\mu > 0$. Consequently, by the strong law of large numbers,

$$\lim_{n \to \infty} S_n/n = \mu > 0,$$

implying that

$$\lim_{n\to\infty} S_n = \infty.$$

Thus, with a probability of one, $S_n < t$ for only a finite number of n, showing that

$$P(N(t) < \infty) = 1$$

and enabling us to write

$$N(t) = \max(n : S_n \le t).$$

The function

$$m(t) = E[N(t)]$$

is called the *renewal function*. We now argue that it is finite for all t.

Proposition 6.1

$$m(t) < \infty.$$

Proof Because $P(X_i \le 0) < 1$, it follows from the continuity property of probabilities that there is a value $\beta > 0$ such that $P(X_i \ge \beta) > 0$. Let

$$\bar{X}_i = \beta\, I_{\{X_i \ge \beta\}},$$

and define the renewal process

$$\bar{N}(t) = \max(n : \bar{X}_1 + \cdots + \bar{X}_n \le t).$$

Because renewals of this process can only occur at integral multiples of β, and because the number of them that occur at the time $n\beta$ is a geometric random variable with parameter $P(X_i \ge \beta)$, it follows that

$$E[\bar{N}(t)] \le \frac{t/\beta + 1}{P(X_i \ge \beta)} < \infty.$$

Because $\bar{X}_i \le X_i, i \ge 1$, implies that $N(t) \le \bar{N}(t)$, the result is proven. ∎

6.2   Some Limit Theorems of Renewal Theory

In this section, we prove the strong law and the central limit theorem for renewal processes as well as the elementary renewal theorem. We start with the strong law for renewal processes, which says that $N(t)/t$ converges almost surely to the inverse of the mean interevent time.

Proposition 6.2 *Strong law for renewal processes.   With a probability of one,*

$$\lim_{t\to\infty} \frac{N(t)}{t} = \frac{1}{\mu} \qquad \left(\text{where } \frac{1}{\infty} \equiv 0\right).$$

Proof Because S_n is the time of the nth event, and $N(t)$ is the number of events by time t, it follows that $S_{N(t)}$ and $S_{N(t)+1}$ represent, respectively, the time of the last event prior to or at time t and the time of the first event after t. Consequently,

$$S_{N(t)} \le t < S_{N(t)+1},$$

implying that

$$\frac{S_{N(t)}}{N(t)} \le \frac{t}{N(t)} < \frac{S_{N(t)+1}}{N(t)}. \tag{6.1}$$

Because $N(t) \to_{as} \infty$ as $t \to \infty$, it follows by the strong law of large numbers that

$$\frac{S_{N(t)}}{N(t)} = \frac{X_1 + \cdots + X_{N(t)}}{N(t)} \to_{as} \mu \quad \text{as } t \to \infty.$$

Similarly,

$$\frac{S_{N(t)+1}}{N(t)} = \frac{X_1 + \cdots + X_{N(t)+1}}{N(t)+1} \frac{N(t)+1}{N(t)} \to_{as} \mu \quad \text{as } t \to \infty,$$

and the result follows. ∎

Example 6.3 Suppose that a coin selected from a bin will on each flip come up heads with a fixed but unknown probability with a probability distribution that is uniformly distributed on $(0,1)$. At any time, the coin currently in use can be discarded and a new coin chosen. The heads probability of this new coin, independent of what has previously transpired, will also have a uniform $(0,1)$ distribution. If the objective is to maximize the long-run proportion of flips that land heads, what is a good strategy?

Solution Consider the strategy of discarding the current coin whenever it lands on tails. Under this strategy, every time a tail occurs we have a renewal. Thus, by the strong law for renewal processes, the long-run proportion of flips that land tails is the inverse of μ, the mean number of flips until a selected coin comes up tails. Because

$$\mu = \int_0^1 \frac{1}{1-p} dp = \infty,$$

it follows that, under this strategy, the long-run proportion of coin flips that come up heads is one. ∎

The elementary renewal theorem says the $E[N(t)/t]$ also converges to $1/\mu$. Before proving it, we will prove a lemma.

Lemma 6.4 *Wald's equation. Suppose that $X_n \geq 1$ are iid with finite mean $E[X]$ and that N is a stopping time for this sequence, in the sense that the event $\{N > n-1\}$ is independent of $X_n, X_{n+1}, \ldots,$ for all n. If $E[N] < \infty$, then*

$$E\left[\sum_{n=1}^{N} X_i\right] = E[N]E[X].$$

Proof To begin, let us prove the lemma when the X_i are replaced by their absolute values. In this case,

$$
\begin{aligned}
E\left[\sum_{n=1}^{N} |X_n|\right] &= E\left[\sum_{n=1}^{\infty} |X_n| I_{\{N \geq n\}}\right] \\
&= E\left[\lim_{m \to \infty} \sum_{n=1}^{m} |X_n| I_{\{N \geq n\}}\right] \\
&= \lim_{m \to \infty} E\left[\sum_{n=1}^{m} |X_n| I_{\{N \geq n\}}\right],
\end{aligned}
$$

where the monotone convergence theorem (Theorem 1.43) was used to justify the interchange of the limit and expectations operations in the last equality. Continuing, we then get

$$
\begin{aligned}
E\left[\sum_{n=1}^{N} |X_n|\right] &= \lim_{m \to \infty} \sum_{n=1}^{m} E[|X_n| I_{\{N > n-1\}}] \\
&= \sum_{n=1}^{\infty} E[|X_n|] E[I_{\{N > n-1\}}] \\
&= E[|X|] \sum_{n=1}^{\infty} P(N > n - 1) \\
&= E[|X|] E[N] \\
&< \infty.
\end{aligned}
$$

But now we can repeat exactly the same sequence of steps, but with X_i replacing $|X_i|$, and with the justification of the interchange of the expectation and limit operations in the third equality now provided by the dominated convergence theorem (Theorem 1.38) upon using the bound $|\sum_{n=1}^{m} X_n I_{\{N \geq n\}}| \leq \sum_{n=1}^{N} |X_i|$. ∎

Proposition 6.5 *Elementary renewal theorem.*

$$\lim_{t \to \infty} \frac{m(t)}{t} = \frac{1}{\mu} \quad \left(where \ \frac{1}{\infty} \equiv 0\right).$$

Proof Suppose first that $\mu < \infty$. Because

$$N(t) + 1 = \min(n : S_n > t),$$

it follows that $N(t) + 1$ is a stopping time for the sequence of interevent times $X_1, X_2, \ldots$. Consequently, by Wald's equation, we see that

$$E[S_{N(t)+1}] = \mu[m(t) + 1].$$

Because $S_{N(t)+1} > t$, the preceding implies that

$$\liminf_{t \to \infty} \frac{m(t)}{t} \geq \frac{1}{\mu}.$$

We will complete the proof by showing that $\limsup_{t \to \infty} \frac{m(t)}{t} \leq \frac{1}{\mu}$. Toward this end, fix a positive constant M and define a related renewal process with interevent times $\bar{X}_n$, $n \geq 1$, given by

$$\bar{X}_n = \min(X_n, M).$$

Let

$$\bar{S}_n = \sum_{i=1}^{n} \bar{X}_i, \qquad \bar{N}(t) = \max(n : \bar{S}_n \leq t).$$

Because an interevent time of this related renewal process is at most M, it follows that

$$\bar{S}_{\bar{N}(t+1)} \leq t + M.$$

Taking expectations and using Wald's equation yields

$$\mu_M[\bar{m}(t) + 1] \leq t + M,$$

where $\bar{\mu}_M = E[\bar{X}_n]$ and $\bar{m}(t) = E[\bar{N}(t)]$. The preceding equation implies that

$$\limsup_{t \to \infty} \frac{\bar{m}(t)}{t} \leq \frac{1}{\mu_M}.$$

However, $\bar{X}_n \leq X_n$, $n \geq 1$, implies that $\bar{N}(t) \geq N(t)$ and thus that $\bar{m}(t) \geq m(t)$. Thus,

$$\limsup_{t \to \infty} \frac{m(t)}{t} \leq \frac{1}{\mu_M}. \tag{6.2}$$

Now,

$$\min(X_1, M) \uparrow X_1 \quad \text{as} \quad M \uparrow \infty,$$

so by the dominated convergence theorem, it follows that

$$\bar{\mu}_M \to \mu \quad \text{as} \quad M \to \infty.$$

Thus, letting $M \to \infty$ in Equation 6.2 yields

$$\limsup_{t \to \infty} \frac{m(t)}{t} \leq \frac{1}{\mu}.$$

Thus, the result is established when $\mu < \infty$. When $\mu = \infty$, again consider the related renewal process with interarrivals $\min(X_n, M)$. Using the monotone convergence theorem, we can conclude that $\mu_M = E[\min(X_1, M)] \to \infty$ as $M \to \infty$. Consequently, Equation 6.2 implies that

$$\limsup \frac{m(t)}{t} = 0,$$

and the proof is complete. ∎

If the interarrival times $X_i, i \geq 1$, of the counting process $N(t), t \geq 0$, are independent, but with X_1 having distribution G, and the other X_i having distribution F, the counting process is said to be a *delayed renewal process*. We leave it as an exercise to show that the analogs of the the strong law and the elementary renewal theorem remain valid.

Remark 6.6 Consider an irreducible recurrent Markov chain. For any state j, we can consider transitions into state j as constituting renewals. If $X_0 = j$, then $N_j(n), n \geq 0$, would be a renewal process, where $N_j(n)$ is the number of transitions into state j by time n; if $X_0 \neq j$, then $N_j(n), n \geq 0$ would be a delayed renewal process. The strong law for renewal processes then shows that, with a probability of one, the long-run proportion of transitions that are into state j is $1/\mu_j$, where μ_j is the mean number of transitions between successive visits to state j. Thus, for positive recurrent irreducible chains the stationary probabilities will equal these long-run proportions of time that the chain spends in each state.

Proposition 6.7 *Central limit theorem for renewal processes. If μ and σ^2, assumed finite, are the mean and variance of an interevent time, then $N(t)$ is asymptotically normal with mean t/μ and variance $t\sigma^2/\mu^3$. That is,*

$$\lim_{t \to \infty} P\left(\frac{Nt) - t/\mu}{\sigma\sqrt{t/\mu^3}} < y\right) = \frac{1}{\sqrt{2\pi}} \int_{-\infty}^{y} e^{-x^2/2} dx.$$

Proof Let $r_t = t/\mu + y\sigma\sqrt{t/\mu^3}$. If r_t is an integer, let $n_t = r_t$; if r_t is not an integer, let $n_t = [r_t] + 1$, where $[x]$ is the largest integer less than or equal to x. Then

$$P\left(\frac{N(t) - t/\mu}{\sigma\sqrt{t/\mu3}} < y\right) = P(N(t) < r_t)$$

$$= P(N(t) < n_t)$$

$$= P(S_{n_t} > t)$$

$$= P\left(\frac{S_{n_t} - n_t\mu}{\sigma\sqrt{n_t}} > \frac{t - n_t\mu}{\sigma\sqrt{n_t}}\right),$$

where the preceding used that the events $\{N(t) < n\}$ and $\{S_n > t\}$ are equivalent. Now, by the central limit theorem, $\frac{S_{n_t} - n_t\mu}{\sigma\sqrt{n_t}}$ converges to a standard normal random variable as n_t approaches ∞ or, equivalently, as t approaches ∞. Also,

$$\lim_{t\to\infty} \frac{t - n_t\mu}{\sigma\sqrt{n_t}} = \lim_{t\to\infty} \frac{t - r_t\mu}{\sigma\sqrt{r_t}}$$

$$= \lim_{t\to\infty} \frac{-y\mu\sqrt{t/\mu^3}}{\sqrt{t/\mu + y\sigma\sqrt{t/\mu^3}}}$$

$$= -y.$$

Consequently, with Z being a standard normal random variable

$$\lim_{t\to\infty} P\left(\frac{Nt - t/\mu}{\sigma\sqrt{t/\mu3}} < y\right) = P(Z > -y) = P(Z < y),$$

and the proof is complete. ∎

6.3   Renewal Reward Processes

Consider a renewal process with interarrival times $X_n, n \geq 1$, and suppose that rewards are earned in such a manner that if R_n is the reward earned during the nth renewal cycle – that is, during the time from S_{n-1} to S_n – then the random vectors (X_n, R_n) are iid. The idea of this definition is that the reward earned during a renewal cycle is allowed to depend on what occurs during that cycle and thus on its length, but whenever a renewal occurs, the process probabilistically restarts. Let $R(t)$ denote the total reward earned by time t.

Theorem 6.8 *If $E[R_1]$ and $E[X_1]$ are both finite, then*

$$(a) \quad \frac{R(t)}{t} \xrightarrow{as} \frac{E[R_1]}{E[X_1]} \quad as \quad t \to \infty$$

$$(b) \quad \frac{E[R(t)]}{t} \to \frac{E[R_1]}{E[X_1]} \quad as \quad t \to \infty.$$

Proof To begin, let us suppose that the reward received during a renewal cycle is earned at the end of that cycle. Consequently,

$$R(t) = \sum_{n=1}^{N(t)} R_n,$$

and thus

$$\frac{R(t)}{t} = \frac{\sum_{n=1}^{N(t)} R_n}{N(t)} \frac{N(t)}{t}.$$

Because $N(t) \to \infty$ as $t \to \infty$, it follows from the strong law of large numbers that

$$\frac{\sum_{n=1}^{N(t)} R_n}{N(t)} \longrightarrow_{as} E[R_1].$$

Hence, Part (a) follows by the strong law for renewal processes.

To prove Part (b), fix $0 < M < \infty$, set $\bar{R}_i = \min(R_i, M)$, and let $\bar{R}(t) = \sum_{i=1}^{N(t)} \bar{R}_i$. Then,

$$E[R(t)] \geq E[\bar{R}(t)]$$

$$= E\left[\sum_{i=1}^{N(t)+1} \bar{R}_i\right] - E[\bar{R}_{N(t)+1}]$$

$$= [m(t) + 1]E[\bar{R}_1] - E[\bar{R}_{N(t)+1}],$$

where the final equality used Wald's equation. Because $\bar{R}_{N(t)+1} \leq M$, the preceding yields

$$\frac{E[R(t)]}{t} \geq \frac{m(t) + 1}{t} E[\bar{R}_1] - \frac{M}{t}.$$

Consequently, by the elementary renewal theorem,

$$\lim_{t \to \infty} \inf \frac{E[R(t)]}{t} \geq \frac{E[\bar{R}_1]}{E[X]}.$$

By the dominated convergence theorem, $\lim_{M \to \infty} E[\bar{R}_1] = E[R_1]$, yielding

$$\lim_{t \to \infty} \inf \frac{E[R(t)]}{t} \geq \frac{E[R_1]}{E[X]}.$$

Letting $R^*(t) = -R(t) = \sum_{i=1}^{N(t)} (-R_i)$ yields, upon repeating the same argument,

$$\lim_{t \to \infty} \inf \frac{E[R^*(t)]}{t} \geq \frac{E[-R_1]}{E[X]}$$

or equivalently,

$$\limsup_{t\to\infty} \frac{E[R(t)]}{t} \le \frac{E[R_1]}{E[X]}.$$

Thus,

$$\lim_{t\to\infty} \frac{E[\sum_{i=1}^{N(t)} R_i]}{t} = \frac{E[R_1]}{E[X]}, \tag{6.3}$$

proving the theorem when the entirety of the reward earned during a renewal cycle is gained at the end of the cycle. Before proving the result without this restriction, note that

$$E\left[\sum_{i=1}^{N(t)} R_i\right] = E\left[\sum_{i=1}^{N(t)+1} R_i\right] - E[R_{N(t)+1}]$$

$$= E[R_1]E[N(t)+1] - E[R_{N(t)+1}] \quad \text{by Wald's equation}$$

$$= E[R_1][m(t)+1] - E[R_{N(t)+1}].$$

Hence,

$$\frac{E[\sum_{i=1}^{N(t)} R_i]}{t} = \frac{m(t)+1}{t} E[R_1] - \frac{E[R_{N(t)+1}]}{t},$$

so we can conclude from Equation 6.3 and the elementary renewal theorem that

$$\frac{E[R_{N(t)+1}]}{t} \to 0. \tag{6.4}$$

Now, let us drop the assumption that the rewards are earned only at the end of renewal cycles. Suppose first that all partial returns are nonnegative. Then, with $R(t)$ equal to the total reward earned by time t,

$$\frac{\sum_{i=1}^{N(t)} R_i}{t} \le \frac{R(t)}{t} \le \frac{\sum_{i=1}^{N(t)} R_i}{t} + \frac{R_{N(t)+1}}{t}.$$

Taking expectations, and using Equations 6.3 and 6.4 proves Part (b). Part (a) follows from the inequality

$$\frac{\sum_{i=1}^{N(t)} R_i}{N(t)} \frac{N(t)}{t} \le \frac{R(t)}{t} \le \frac{\sum_{i=1}^{N(t)+1} R_i}{N(t)+1} \frac{N(t)+1}{t}$$

by noting that for $j = 0, 1$

$$\frac{\sum_{i=1}^{N(t)+j} R_i}{N(t)+j} \frac{N(t)+j}{t} \xrightarrow{as} \frac{E[R_1]}{E[X_1]}.$$

A similar argument holds when all partial returns are nonpositive, and the general case follows by breaking up the returns into their positive and negative parts and applying the preceding argument separately to each. ∎

Example 6.9 *Generating a random variable with a distribution that is the stationary distribution of a Markov chain.* For a finite-state irreducible aperiodic Markov chain $X_n, n \geq 0$, having transition probabilities $\{P_{ij}\}$ and stationary distribution $\{\pi_i\}$, Theorem 5.13 says that the approximation $P(X_n = i|X_0 = 0) \approx \pi_i$ is good for large n. Here we will show how to find a random time $T \geq 0$ so that we have exactly $P(X_T = i|X_0 = 0) = \pi_i$.

Suppose that for some $p > 0$ we have $P_{i0} > p$ for all i. (If this condition doesn't hold, then we can always find an m such that the condition holds for the transition probabilities $P_{ij}^{(m)}$, implying that the condition holds for the Markov chain $Y_n = X_{nm}, n \geq 0$, which also has the stationary distribution $\{\pi_i\}$.)

To begin, let $J_n \sim \text{Bernoulli}(p)$ be iid and define a Markov chain Y_n so that

$$P(Y_{n+1} = 0|Y_n = i, J_{n+1} = 1) = 1,$$

$$P(Y_{n+1} = 0|Y_n = i, J_{n+1} = 0) = (P_{i0} - p)/(1 - p),$$

and for $j \neq 0$,

$$P(Y_{n+1} = j|Y_n = i, J_{n+1} = 0) = P_{ij}/(1 - p).$$

Notice that this gives $Y_n = 0$ whenever $J_n = 1$ and in addition that

$$
\begin{aligned}
P(Y_{n+1} = j|Y_n = i) &= P(Y_{n+1} = j|Y_n = i, J_{n+1} = 0)(1 - p) \\
&\quad + P(Y_{n+1} = j|Y_n = i, J_{n+1} = 1)p \\
&= P_{ij}.
\end{aligned}
$$

Thus, both X_n and Y_n have the same transition probabilities and thus the same stationary distribution.

Say that a new cycle begins at time n if $J_n = 1$. Suppose that a new cycle begins at time zero, so $Y_0 = 0$, and let N_j denote the number of time periods the chain is in state j during the first cycle. If we suppose that a reward of one is earned each time the chain is in state j, then π_j equals the long-run average reward per unit time, and the renewal reward process result yields that

$$\pi_j = \frac{E[N_j]}{E[T]},$$

where $T = \min\{n > 0 : J_n = 1\}$ is the time of the first cycle. Because T is geometric with parameter p, we obtain the identity

$$\pi_j = pE[N_j].$$

Now, let I_k be the indicator variable for the event that a new cycle begins on the transition following the kth visit to state j. Note that

$$\sum_{k=1}^{N_j} I_k = I_{\{Y_{T-1}=j\}}.$$

Because $I_1, I_2, \ldots$ are iid and the event $\{N_j = n\}$ is independent of I_{n+1}, $I_{n+2}, \ldots$, it follows from Wald's equation that

$$P(Y_{T-1} = j) = E[N_j]E[I_1] = pE[N_j],$$

giving the result that

$$\pi_j = P(Y_{T-1} = j).$$

Remark 6.10 In terms of the original Markov chain X_n, set $X_0 = 0$. Let $U_1, U_2, \ldots$ be a sequence of independent uniform $(0, 1)$ random variables that is independent of the Markov chain. Then define

$$T = \min(n > 0 : X_n = 0, U_n < p/P_{X_{n-1},0}),$$

with the result that $P(X_{T-1} = j | X_0 = 0) = \pi_j$. In fact, if we set $X_0 = 0$, let $T_0 = 0$, and define

$$T_i = \min(n > T_{i-1} : X_n = 0, U_n < p/P_{X_{n-1},0}),$$

then $X_{T_i-1}, i \geq 1$, are iid with

$$P(X_{T_i-1} = j | X_0 = 0) = \pi_j.$$

Example 6.11 Suppose that $X_i, i \geq 1$ are iid discrete random variables with probability mass function $p_i = P(X = i)$. Suppose we want to find the expected time until the pattern $1, 2, 1, 3, 1, 2, 1$ appears. To do so, suppose that we earn a reward of one each time the pattern occurs. Because a reward of one is earned at time $n \geq 7$ with probability $P(X_n = 1, X_{n-1} = 2, X_{n-2} = 1, \ldots, X_{n-6} = 1) = p_1^4 p_2^2 p_3$, it follows that the long-run expected reward per unit time is $p_1^4 p_2^2 p_3$. However, suppose that the pattern has just occurred at time 0. Say that cycle 1 begins at time 1, and that a cycle ends when, ignoring data from previous cycles, the pattern reappears. Thus, for instance, if a cycle has just ended then the last data value was 1, the next to last was 2, then 1, then 3, then 1, then 2, and then 1. The next cycle will begin when, without using any of these values, the pattern reappears. The total reward earned during a cycle, call it R, can be expressed as

$$R = 1 + A_4 + A_6,$$

where A_4 is the reward earned when we observe the fourth data value of the cycle (it will equal one if the first four values in the cycle are $3, 1, 2, 1$), A_6 is the reward earned when we observe the sixth data value of the cycle, and one is the reward earned when the cycle ends. Hence,

$$E[R] = 1 + p_1^2 p_2 p_3 + p_1^3 p_2^2 p_3.$$

If T is the time of a cycle, then by the renewal reward theorem,

$$p_1^4 p_2^2 p_3 = \frac{E[R]}{E[T]},$$

yielding that the expected time until the pattern appears is

$$E[T] = \frac{1}{p_1^4 p_2^2 p_3} + \frac{1}{p_1^2 p_2} + \frac{1}{p_1}.$$

Let

$$A(t) = t - S_{N(t)}, \qquad Y(t) = S_{N(t)+1} - t.$$

The random variable $A(t)$, equal to the time at t since the last renewal prior to (or at) time t, is called the *age* of the renewal process at t. The random variable $Y(t)$, equal to the time from t until the next renewal, is called the *excess* of the renewal process at t. We now apply renewal reward processes to obtain the long-run average values of the age and of the excess as well as the long-run proportions of time that the age and the excess are less than x.

The distribution function F_e defined by

$$F_e(x) = \frac{1}{\mu} \int_0^x \bar{F}(y)\, dy, \quad x \geq 0$$

is called the *equilibrium* distribution of the renewal process.

Proposition 6.12 *Let X have distribution F. With a probability of one,*

(a) $\displaystyle \lim_{t \to \infty} \frac{1}{t} \int_0^t A(s)\, ds = \lim_{t \to \infty} \frac{1}{t} \int_0^t E[A(s)]\, ds = \frac{E[X^2]}{2\mu}$

(b) $\displaystyle \lim_{t \to \infty} \frac{1}{t} \int_0^t Y(s)\, ds = \lim_{t \to \infty} \frac{1}{t} \int_0^t E[Y(s)]\, ds = \frac{E[X^2]}{2\mu}$

(c) $\displaystyle \lim_{t \to \infty} \frac{1}{t} \int_0^t I_{\{A(s) < x\}}\, ds = \lim_{t \to \infty} \frac{1}{t} \int_0^t P(A(s) < x)\, ds = F_e(x)$

(d) $\displaystyle \lim_{t \to \infty} \frac{1}{t} \int_0^t I_{\{Y(s) < x\}}\, ds = \lim_{t \to \infty} \frac{1}{t} \int_0^t P(Y(s) < x)\, ds = F_e(x).$

Proof To prove Part (a), imagine that a reward at rate $A(s)$ is earned at time s, $s \geq 0$. Then, this reward process is a renewal reward process with a new cycle beginning each time a renewal occurs. Because the reward rate at a time x units into a cycle is x, it follows that if X is the length of a cycle, then T, the total reward earned during a cycle, is

$$T = \int_0^X x\, dx = X^2/2.$$

Because $\lim_{t\to\infty} \frac{1}{t} \int_0^t A(s)ds$ is the long run average reward per unit time, Part (a) follows from the renewal reward theorem.

To prove Part (c) imagine that we earn a reward at a rate of one per unit time whenever the age of the renewal process is less than x. That is, at time s we earn a reward at rate one if $A(s) < x$ and at rate zero if $A(s) > x$. Then this reward process is also a renewal reward process in which a new cycle begins whenever a renewal occurs. Because we earn at rate one during the first x units of a renewal cycle and at rate zero for the remainder of the cycle, it follows that, with T being the total reward earned during a cycle,

$$
\begin{aligned}
E[T] &= E[\min(X, x)] \\
&= \int_0^\infty P(\min(X, x) > t)\, dt \\
&= \int_0^x \bar{F}(t)dt.
\end{aligned}
$$

Because $\lim_{t\to\infty} \frac{1}{t} \int_0^t I_{\{A(s)<x\}}\, ds$ is the average reward per unit time, Part (c) follows from the renewal reward theorem.

We will leave the proofs of Parts (b) and (d) as an exercise.

6.4 Queuing Theory Applications of Renewal Reward Processes

Suppose that customers arrive to a system according to a renewal process having interarrival distribution F, with mean $1/\lambda$. Each arriving customer is eventually served and departs the system. Suppose that at each time point the system is in some state, and let $S(t)$ denote the state of the system at time t. Suppose that when an arrival finds the system empty of other customers the evolution of system states from this point on is independent of the past and has the same distribution each time this event occurs. (We often say that the state process probabilistically restarts every time an arrival finds the system empty.) Suppose that such an event occurs at time zero.

If we suppose that each arrival pays an amount to the system, with that amount being a function of the state history while that customer is in the system, then the resulting reward process is a renewal reward process, with a new cycle beginning each time an arrival finds the system empty of other customers. Hence, if $R(t)$ denotes the total reward earned by time t, and T denotes the length of a cycle, then

$$
\text{average reward per unit time} = \frac{E[R(T)]}{E[T]}. \tag{6.5}
$$

Now, let R_i denote the amount of money paid by customer i, for $i \geq 1$, and let N denote the number of customers served in a cycle. Then this sequence $R_1, R_2, \ldots$ can be thought of as the reward sequence of a renewal reward process in which R_i is the reward earned during period i and the cycle time is N. Hence, by the renewal reward process result,

$$\lim_{n \to \infty} \frac{R_1 + \cdots + R_n}{n} = \frac{E[\sum_{i=1}^{N} R_i]}{E[N]}. \tag{6.6}$$

To relate the left-hand sides of Equations 6.5 and 6.6, first note that because $R(T)$ and $\sum_{i=1}^{N} R_i$ both represent the total reward earned in a cycle, they must be equal. Thus,

$$E[R(T)] = E\left[\sum_{i=1}^{N} R_i \right].$$

Also, with customer 1 being one who found the system empty at the arrival time zero, if we let X_i denote the time between the arrivals of customers i and $i+1$, then

$$T = \sum_{i=1}^{N} X_i.$$

Because the event $\{N = n\}$ is independent of all the sequence $X_k, k \geq n+1$, it follows that it is a stopping time for the sequence $X_i, i \geq 1$. Thus, Wald's equation gives that

$$E[T] = E[N]E[X] = \frac{E[N]}{\lambda},$$

and we have proven the following.

Proposition 6.13

$$\textit{average reward per unit time} = \lambda \bar{R},$$

where

$$\bar{R} = \lim_{n \to \infty} \frac{R_1 + \cdots + R_n}{n}$$

is the average amount that a customer pays.

Corollary 6.14 *Let $X(t)$ denote the number of customers in the system at time t, and set*

$$L = \lim_{t \to \infty} \frac{1}{t} \int_0^t X(s)ds.$$

Also, let W_i denote the amount of time that customer i spends in the system, and set

$$W = \lim_{n \to \infty} \frac{W_1 + \cdots + W_n}{n}.$$

Then, the preceding limits exist, L and W are both constants, and

$$L = \lambda W.$$

Proof Imagine that each customer pays one per unit time while in the system. Then the average reward earned by the system per unit time is L, and the average amount a customer pays is W. Consequently, the result follows directly from Proposition 6.13. ∎

6.5 Blackwell's Theorem

A discrete interarrival distribution F is said to be *lattice* with period d if $\sum_{n \geq 0} P(X_i = nd) = 1$ and d is the largest value having this property. (Not every discrete distribution is lattice. For instance, the two point distribution that puts all its weights on the values one and π, or any other irrational number, is not lattice.) In this case, renewals can only occur at integral multiples of d. By letting d be the new unit of time, we can reduce any lattice renewal process to one with interarrival times that put all their weight on the nonnegative integers and are such that the greatest common divisor of $\{n : P(X_i = n) > 0\}$ is one. (If μ' was the original mean interarrival time then in terms of our new units, the new mean μ would equal μ'/d.) So let us suppose that the interarrival distribution is lattice with period one, let $p_j = P(X_i = j), j \geq 0$, and $\mu = \sum_j j p_j$.

Theorem 6.15 *Blackwell's theorem. If the interarrival distribution is lattice with period one, then*

$$\lim_{n \to \infty} P(a \ renewal \ occurs \ at \ time \ n) = \frac{1 - p_0}{\mu}$$

and

$$\lim_{n \to \infty} E[number \ of \ renewals \ at \ time \ n] = \frac{1}{\mu}.$$

Proof With A_n equal to the age of the renewal process at time n, it is easy to see that $A_n, n \geq 0$, is an irreducible, aperiodic Markov chain with transition probabilities

$$P_{i,0} = P(X = i | X \geq i) = \frac{p_i}{\sum_{j \geq i} p_j} = 1 - P_{i,i+1}, \ i \geq 0.$$

The limiting probability that this chain is in state zero is

$$\pi_0 = \frac{1}{E[X | X > 0]}.$$

where X is an interarrival time of the renewal process. (The mean number of transitions of the Markov chain between successive visits to state zero is $E[X | X > 0]$ because an interarrival that is equal to zero is ignored by the chain.) Because a renewal occurs at time n whenever $A_n = 0$, the first part of the theorem is proven. The second part of the theorem also follows because, conditional on a renewal occurring at time n, the total number of renewals that occur at that time is geometric with mean $(1 - p_0)^{-1}$. ∎

6.6 Poisson Process

If the interarrival distribution of a renewal process is exponential with rate λ, then the renewal process is said to be a Poisson process with rate λ. Why it is called a Poisson process is answered by the next proposition.

Proposition 6.16 *If $N(t), t \geq 0$, is a Poisson process having rate λ, then $N(t) =_d$ Poisson(λt).*

Proof We will show $P(N(t) = k) = e^{-\lambda t}(\lambda t)^k/k!$ by induction on k. Note first that $P(N(t) = 0) = P(X_1 > t) = e^{-\lambda t}$. For $k > 0$, we condition on X_1 to get

$$
\begin{aligned}
P(N(t) = k) &= \int_0^\infty P(N(t) = k | X_1 = x)\lambda e^{-\lambda x} dx \\
&= \int_0^t P(N(t - x) = k - 1)\lambda e^{-\lambda x} dx \\
&= \int_0^t \frac{e^{-\lambda(t-x)}(\lambda(t - x))^{k-1}}{(k - 1)!}\lambda e^{-\lambda x} dx \\
&= \lambda^k e^{-\lambda t}\int_0^t \frac{(t - x)^{k-1}}{(k - 1)!}dx \\
&= e^{-\lambda t}(\lambda t)^k/k!,
\end{aligned}
$$

which completes the induction proof. ∎

The Poisson process is often used as a model of customer arrivals to a queuing system because the process has several properties that might be expected of such a customer arrival process. The process $N(t), t \geq 0$, is said to be a *counting process* if events are occurring randomly in time and $N(t)$ denotes the cumulative number of such events that occur in $[0, t]$. The counting process $N(t), t \geq 0$, is said to have *stationary increments* if the distribution of $N(t + s) - N(t)$ does not depend on t and is said to have *independent increments* if $N(t_i + s_i) - N(t_i)$, for $i = 1, 2, \ldots$, are independent random variables whenever $t_{i+1} > t_i + s_i$ for all i. Independent increments say that customer traffic in one interval of time does not affect traffic in another (disjoint) interval of time. Stationary increments say that the traffic process is not changing over time. Our next proposition shows that the Poisson process is the only possible counting process with continuous interarrival times and with both of these properties.

Proposition 6.17 *The Poisson process is the only counting process with stationary, independent increments and continuous interarrival times.*

Proof Given a counting process $N(t), t \geq 0$, with continuous interarrival times X_i and stationary, independent increments, then

$$\begin{aligned}
P(X_1 > t + s | X_1 > t) &= P(N(t + s) = 0 | N(t) = 0) \\
&= P(N(t + s) - N(t) = 0 | N(t) = 0) \\
&= P(N(t + s) - N(t) = 0) \\
&= P(N(s) = 0) \\
&= P(X_1 > s),
\end{aligned}$$

where the third equality follows from independent increments and the fourth from stationary increments. Thus, we see that X_1 is memoryless. Because the only memoryless continuous random variable is the exponential distribution, X_1 is exponential. Because

$$\begin{aligned}
P(X_2 > s | X_1 = t) &= P(N(t + s) - N(t) = 0 | X_1 = t) \\
&= P(N(t + s) - N(t) = 0) \\
&= P(N(s) = 0) \\
&= P(X_1 > s),
\end{aligned}$$

it follows that X_2 is independent of X_1 and has the same distribution. Continuing in this way shows that all the X_i are iid exponentials, so $N(t), t \geq 0$, is a Poisson process. ∎

6.7   Exercises

1. Consider a renewal process $N(t)$ with Bernoulli(p) interevent times.
   (a) Compute the distribution of $N(t)$.
   (b) With $S_i, i \geq 1$ equal to the time of event i, find the conditional probability mass function of $S_1, \ldots, S_k$ given that $N(n) = k$.

2. For a renewal process with an interevent distribution F with density $F' = f$, prove the renewal equation

$$m(t) = F(t) + \int_0^t m(t - x) f(x) dx.$$

3. For a renewal process with an interevent distribution F, show that

$$P(X_{N(t)+1} > x) \geq 1 - F(x).$$

The preceding states that the length of the renewal interval that contains the point t is stochastically larger than an ordinary renewal interval and is called the inspection paradox.

4. With $X_1, X_2, \ldots$ independent $U(0,1)$ random variables with $S_n = \sum_{i \leq n} X_i$ and $N = \min\{n : S_n > 1\}$, show that $E[S_N] = e/2$.

5. A room in a factory has n machines that are always all turned on at the same time, and each works an independent exponential time with mean m days before breaking down. As soon as k machines break, a repairman is called. The repairman takes exactly d days to arrive and instantly repairs all the broken machines. Then this cycle repeats.
   (a) How often in the long run does the repairman get called?
   (b) What is the distribution of the total number of broken machines the repairman finds upon arrival?
   (c) What fraction of time in the long run are there more than k broken machines in the room?

6. Each item produced is either defective or acceptable. Initially, each item is inspected, and this continues until k consecutive acceptable items are discovered. At this point, 100% inspection stops and each new item produced is, independently, inspected with probability α. This continues until a defective item is found, at which point we go back to 100% inspection, with the inspection rule repeating itself. If each item produced is, independently, defective with probability q, what proportion of items are inspected?

7. A system consists of two independent parts, with part i functioning for an exponentially distributed time with rate λ_i before failing, $i = 1, 2$. The system functions as long as at least one of these two parts is working. When the system stops functioning, a new system, with two working parts, is put into use. A cost K is incurred whenever this occurs; also, operating costs at rate c per unit time are incurred whenever the system is operating with both parts working, and operating costs at rate c_i are incurred whenever the system is operating with only part i working, $i = 1, 2$. Find the long-run average cost per unit time.

8. If the interevent times $X_i, i \geq 1$, are independent but with X_1 having a different distribution from the others, then $\{N_d(t), t \geq 0\}$ is called a *delayed renewal process*, where

$$N_d(t) = \sup\left\{n : \sum_{i=1}^{n} X_i \leq t\right\}.$$

Show that the strong law remains valid for a delayed renewal process.

9. Prove Parts (b) and (d) of Proposition 6.12.

10. Consider a renewal process with continuous interevent times $X_1, X_2, \ldots$ having distribution F. Let Y be independent of the X_i and have distribution function F_e. Show that

$$E[\min\{n : X_n > Y\}] = \sup\{x : P(X > x) > 0\}/E[X],$$

where X has distribution F. How can you interpret this result?

11. Someone rolls a die repeatedly and adds up the numbers. Which is larger: P(sum ever hits 2) or P(sum ever hits 102)?

12. If $\{N_i(t), t \geq 0\}$, $i = 1, \ldots, k$ are independent Poisson processes with respective rates $\lambda_i, 1 \leq i \leq k$, show that $\sum_{i=1}^{k} N_i(t), t \geq 0$, is a Poisson process with rate $\lambda = \sum_{i=1}^{k} \lambda_i$.

13. A system consists of one server and no waiting space. Customers who arrive when the server is busy are lost. There are n types of customers: Type i customers arrive according to a Poisson process with rate λ_i and have a service time distribution that is exponential with rate μ_i, with the n Poisson arrival processes and all the service times being independent.
(a) What fraction of time is the server busy?
(b) Let X_n be the type of customer (or zero if no customer) in the system immediately prior to the nth arrival. Is this a Markov chain? Is it time reversible?

14. Let $X_i, i = 1, \ldots, n$ be iid continuous random variables having density function f. Letting

$$X_{(i)} = ith \text{ smallest value of } X_1, \ldots, X_n,$$

the random variables $X_{(1)}, \ldots, X_{(n)}$ are called order statistics. Find their joint density function.

15. Let S_i be the time of event i of a Poisson process with rate λ.
(a) Show that, conditional on $N(t) = n$, the variables $S_1, \ldots, S_n$ are distributed as the order statistics of a set of n iid uniform $(0, t)$ random variables.
(b) If passengers arrive at a bus stop according to a Poisson process with rate λ, and the bus arrives at time t, find the expected sum of the amounts of times that each boarding customer has spent waiting at the stop.

16. Suppose that events occur according to a Poisson process with rate λ and that an event occurring at time s is, independent of what has transpired before time s, classified either as a type 1 or as a type 2 event, with respective probabilities $p_1(s)$ and $p_2(s) = 1 - p_1(s)$. Letting $N_i(t)$ denote the number of type i events by time t, show that $N_1(t)$ and $N_2(t)$ are independent Poisson random variables with means $E[N_i(t)] = \lambda \int_0^t p_i(s)ds$.

7

Brownian Motion

7.1 Introduction

One goal of this chapter is to give one of the most beautiful proofs of the central limit theorem, one which does not involve characteristic functions. To do this, we will give a brief tour of continuous time martingales and Brownian motion and demonstrate how the central limit theorem can be essentially deduced from the fact that Brownian motion is continuous.

In Section 7.2, we introduce continuous time martingales, and in Section 7.3, we demonstrate how to construct Brownian motion, prove it is continuous, and show how the self-similar property of Brownian motion leads to an efficient way of estimating the price of path-dependent stock options using simulation. In Section 7.4, we show how random variables can be embedded in Brownian motion, and in Section 7.5, we use this to prove a version of the central limit theorem for martingales.

7.2 Continuous Time Martingales

Suppose we have sigma fields $\mathcal{F}_t$ indexed by a continuous parameter t so that $\mathcal{F}_s \subseteq \mathcal{F}_t$ for all $s \leq t$.

Definition 7.1 *We say that $X(t)$ is a continuous time martingale for $\mathcal{F}_t$ if for all t and $0 \leq s \leq t$ we have*

1. $E|X(t)| < \infty$

2. $X(t) \in \mathcal{F}_t$

3. $E\left[X(t)|\mathcal{F}_s\right] = X(s)$.

Example 7.2 Let $N(t)$ be a Poisson process having rate λ, and let $\mathcal{F}_t = \sigma(N(s), 0 \le s \le t)$. Then $X(t) = N(t) - \lambda t$ is a continuous time martingale because

$$E[N(t) - \lambda t | \mathcal{F}_s] = E[N(t) - N(s) - \lambda(t - s) | \mathcal{F}_s] + N(s) - \lambda s$$
$$= N(s) - \lambda s.$$

We say a process $X(t)$ has stationary increments if $X(t + s) - X(t) =_d X(s) - X(0)$ for all $t, s \ge 0$. We also say a process $X(t)$ has independent increments if $X(t_1) - X(t_0), X(t_2) - X(t_1), \ldots$ are independent random variables whenever $t_0 \le t_1 \le \cdots$.

Although a Poisson process has stationary independent increments, it does not have continuous sample paths. We will show here that it is possible to construct a process with continuous sample paths, called Brownian motion, which in fact is the only possible martingale with continuous sample paths and stationary and independent increments. These properties will be key in proving the central limit theorem.

7.3   Constructing Brownian Motion

Brownian motion is a continuous time martingale that produces a randomly selected path typically looking something like in Figure 7.1. Here we show how to construct Brownian motion $B(t)$ for $0 \le t \le 1$. To get Brownian motion over a wider interval, you can repeat the construction over more unit intervals each time continuing the path from where it ends in the previous interval.

Given a line segment, we say we "move the midpoint up by the amount z" if we are given a line segment connecting the point (a, b) to the point

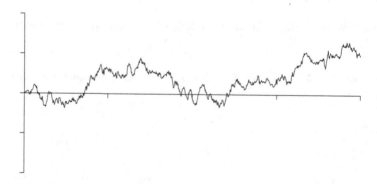

Figure 7.1 **A Brownian motion path.**

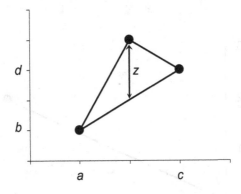

Figure 7.2 **Moving the midpoint of a segment up by** z.

(c, d) with midpoint $(\frac{a+c}{2}, \frac{b+d}{2})$, and we break it into two line segments connecting (a, b) to $(\frac{a+c}{2}, \frac{b+d}{2} + z)$ to (c, d). This is illustrated in Figure 7.2. Next let $Z_{k,n}$ be iid $N(0, 1)$ random variables, for all k, n. We initiate a sequence of paths. The zeroth path consists of the line segment connecting the point $(0, 0)$ to $(1, Z_{0,0})$. For $n \geq 1$, path n will consist of 2^n connected line segments, which can be numbered from left to right. To go from path $n - 1$ to path n, simply move the midpoint of the kth line segment of path $n - 1$ up by the amount $Z_{k,n}/(\sqrt{2})^{n+1}$, $k = 1, \ldots, 2^{n-1}$. Letting $f_n(t)$ be the equation of the nth path, then the random function

$$B(t) = \lim_{n \to \infty} f_n(t)$$

is called standard Brownian motion.

For example, if $Z_{0,0} = 2$ then path zero would be the line segment connecting $(0, 0)$ to $(1, 2)$. This looks like Figure 7.3. Then, if $Z_{1,1}/(\sqrt{2})^2 = 1$, we would move the midpoint $(\frac{1}{2}, 1)$ up to $(\frac{1}{2}, 2)$ and thus path one would consist of the two line segments connecting $(0, 0)$ to $(\frac{1}{2}, 2)$ to $(1, 2)$. This then gives us the path in Figure 7.4.

If $Z_{1,2}/(\sqrt{2})^3 = -1$ and $Z_{2,2}/(\sqrt{2})^3 = 1$, then the next path is obtained by replacing these two line segments with the four line segments connecting $(0, 0)$ to $(\frac{1}{4}, 0)$ to $(\frac{1}{2}, 2)$ to $(\frac{3}{4}, 3)$ to $(1, 2)$. This gives us the path in Figure 7.5. Then the next path would have eight line segments and so on.

Remark 7.3 By this recursive construction, it can immediately be seen that $B(t)$ is "self similar" in the sense that $\{B(t/2^n)(\sqrt{2})^n, 0 \leq t \leq 1\}$ has the same distribution as $\{B(t), 0 \leq t \leq 1\}$. This is the famous fractal property of Brownian motion.

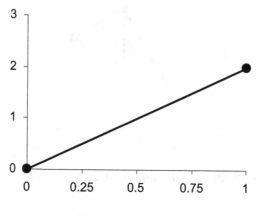

Figure 7.3 **Path 0.**

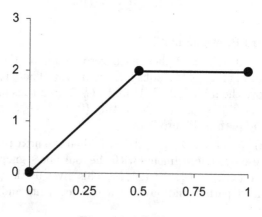

Figure 7.4 **Path 1.**

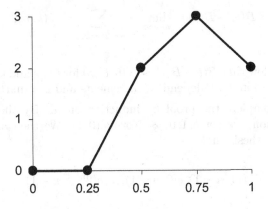

Figure 7.5 **Path 2.**

Proposition 7.4 *Brownian motion $B(t)$ is a martingale with stationary, independent increments and $B(t) \sim N(0,t)$.*

Before we prove this, we need a lemma.

Lemma 7.5 *If X and Y are iid mean zero normal random variables, then the pair $Y + X$ and $Y - X$ are also iid mean zero normal random variables.*

Proof Because X, Y are independent, the pair X, Y has a bivariate normal distribution. Consequently, $X - Y$ and $X + Y$ have a bivariate normal distribution, and thus it's immediate that $Y + X$ and $Y - X$ are identically distributed normal. Then

$$\text{Cov}(Y + X, Y - X) = E[(Y + X)(Y - X)] = E[Y^2 - X^2] = 0$$

gives the result (because uncorrelated bivariate normal random variables are independent). ∎

Proof *Proof of Proposition 7.4.* Letting

$$b(k, n) = B(k/2^n) - B((k - 1)/2^n),$$

we will prove that for any n the random variables

$$b(1, n), b(2, n), ..., b(2^n, n)$$

are iid Normal$(0, 1/2^n)$ random variables. After we prove that Brownian motion is a continuous function, which we do in the proposition immediately following this proof, we can then write

$$B(t) - B(s) = \lim_{n \to \infty} \sum_{k: s + 1/2^n < k/2^n < t} b(k, n).$$

It will then follow that $B(t) - B(s) \sim N(0, t-s)$ for $t > s$ and that Brownian motion has stationary, independent increments and is a martingale.

We will complete the proof by induction on n. By the first step of the construction, we get $b(1, 0) \sim \text{Normal}(0, 1)$. We then assume as our induction hypothesis that

$$b(1, n-1), b(2, n-1), ..., b(2^{n-1}, n-1)$$

are iid $\text{Normal}(0, 1/2^{n-1})$ random variables. Following the rules of the construction, we have

$$b(2k-1, n) = b(k, n-1)/2 + Z_{2k,n}/(\sqrt{2})^{n+1}$$

and

$$\begin{aligned} b(2k, n) &= b(k, n-1) - b(2k-1, n) \\ &= b(k, n-1)/2 - Z_{2k,n}/(\sqrt{2})^{n+1}, \end{aligned}$$

which is also illustrated in Figure 7.6. In the figure, we write $Z = Z_{2k,n}/(\sqrt{2})^{n+1}$. Because $b(k, n-1)/2$ and $Z_{2k,n}/(\sqrt{2})^{n+1}$ are iid Normal $(0, 1/2^{n+1})$ random variables, we then apply the previous lemma to obtain

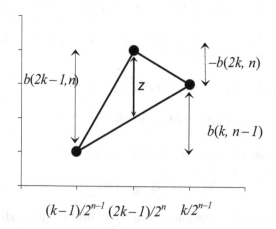

Figure 7.6 **One step of the construction of Brownian motion.**

that $b(2k-1,n)$ and $b(2k,n)$ are iid Normal$(0,1/2^n)$ random variables. Because $b(2k-1,n)$ and $b(2k,n)$ are independent of $b(j,n-1), j \neq k$, we get that

$$b(1,n), b(2,n), ..., b(2^n,n)$$

are iid Normal$(0,1/2^n)$ random variables. ∎

And even though each function $f_n(t)$ is continuous, it is not immediately obvious that the limit $B(t)$ is continuous. For example, if we instead always moved midpoints by a nonrandom amount x, we would have $\sup_{t \in A} B(t) - \inf_{t \in A} B(t) \geq x$ for any interval A, and thus $B(t)$ would not be a continuous function. We next show that Brownian motion is a continuous function.

Proposition 7.6 *Brownian motion is a continuous function with a probability of one.*

Proof Note that

$$P(B(t) \text{ is not continuous})$$

$$\leq \sum_{i=1}^{\infty} P(B(t) \text{ has a discontinuity larger than } 1/i),$$

so the theorem will be proved if we show, for any $\epsilon > 0$,

$$P(B(t) \text{ has a discontinuity larger than } \epsilon) = 0;$$

see the remark immediately following this proof for a discussion of a measurability subtlety.

Because by construction f_m is continuous for any given m, in order for $B(t)$ to have a discontinuity larger than ϵ, we must have

$$\sup_{0 \leq t \leq 1} |B(t) - f_m(t)| > \epsilon/2$$

or else $B(t)$ would necessarily always be within $\epsilon/2$ of the known continuous function f_m, and it would be impossible for $B(t)$ to have a discontinuity larger than ϵ. Letting

$$d_n = \sup_{0 \leq t \leq 1} |f_{n-1}(t) - f_n(t)|$$

be the largest difference between the function at stage n and stage $n+1$, we must then have

$$d_n > \epsilon(3/4)^n/8 \text{ for some } n > m$$

because otherwise it would mean

$$\sup_{0 \le t \le 1} |B(t) - f_m(t)| \le \sum_{n>m} d_n \le \epsilon/2 \sum_{n>m} (3/4)^n/4 \le \epsilon/2.$$

Next note by the construction we have

$$
\begin{aligned}
P(d_n > x) &= P\left(\sup_{1 \le k \le 2^{n-1}} |Z_{k,n}/(\sqrt{2})^{n+1}| > x \right) \\
&\le 2^n P\left(|Z| > (\sqrt{2})^{n+1} x \right) \\
&\le \exp(n - (\sqrt{2})^{n+1} x),
\end{aligned}
$$

where the last line is for sufficiently large n and we use $2^n < e^n$ and $P(|Z| > x) \le e^{-x}$ for sufficiently large x (see Example 4.5).

This together means, for sufficiently large m,

$$P(B(t) \text{ has a discontinuity larger than } \epsilon)$$
$$\le \sum_{n \ge m} P(d_n > \epsilon(3/4)^n/8)$$
$$\le \sum_{n \ge m} \exp(n - \epsilon(3\sqrt{2}/4)^n/8),$$

which because the final sum is finite, can be made arbitrarily close to zero as m increases. ∎

Remark 7.7 The argument in the previous proposition shows that the event

$$\{B(t) \text{ is not continuous}\}$$

is a subset of an event having probability zero. You might now be wondering if this event itself is measurable because the first example in this book shows that there are some seemingly innocent events that are subsets of measurable events but can't have probabilities assigned to them in a consistent way. It turns out this in fact can be a measurable event; the general principle is that any subset of an event having zero probability can be measurable simply by assigning a zero probability to it – and this won't cause any inconsistencies.

For example, consider the set of family heads in the circle as defined in Section 1.2 of this book. Now suppose instead of picking a random point in this circle, you decide to pick a random point in a different circle with no family heads. In this case, the chance you get a family head is zero, and adding up the chances you are i steps away from a family head still adds up to zero without any inconsistency. This means that getting a point in the same family heads set can be made measurable if it is a subset of an event that already has zero probability assigned to it.

The inconsistency in the first example of this book was caused by having a countably infinite number of disjoint events with the same probability that should have probabilities that add up to one, which is impossible. If the probabilities should add up to zero, as would be the case with subsets of an event having probability zero, we can just assign a zero probability to all such events without any inconsistencies. This is made rigorous, with some additional reasoning, by adding all subsets of zero probability events into the initial collection of events that generate the sigma field.

Remark 7.8 You may notice that the only property of the standard normal random variable Z used in the proof is that $P(|Z| \geq x) \leq e^{-x}$ for sufficiently large x. This means we could have instead constructed a process starting with $Z_{k,n}$ having an exponential distribution, and we would get a different limiting process with continuous paths. We would not, however, have stationary and independent increments.

As it does for Markov chains, the strong Markov property holds for Brownian motion. This means that $\{B(T+t) - B(T), 0 \leq t\}$ has the same distribution as $\{B(t), 0 \leq t\}$ for finite stopping times $T < \infty$. We leave a proof of this as Exercise 8 at the end of the chapter. An easy result using the continuity of Brownian motion and the strong Markov property for Brownian motion involves the suprememum of Brownian motion.

Proposition 7.9 $P(\sup_{0 \leq s \leq t} B(s) > x) = 2P(B(t) > x)$.

Proof Let $T = \inf\{t \geq 0 : B(t) = x\}$, and note that continuity of Brownian motion gives the first line here:

$$P(B(t) > x) = P(B(t) > x, T < t)$$
$$= P(T < t)P(B(t) - B(T) > 0 | T < t)$$
$$= P(T < t)/2$$
$$= P\left(\sup_{0 \leq s \leq t} B(s) > x\right)/2,$$

Also note the strong Markov property gives the third line. ∎

Example 7.10 *Path-dependent stock options.* It is most common that the payoff from exercising a stock option depends on the price of a stock at a fixed point in time. If it depends on the price of a stock at several points in time, it is usually called a *path-dependent option*. Although there are many formulas for estimating the value of various different types of stock options, many path-dependent options are commonly valued using Monte

Carlo simulation. Here we give an efficient way to do this using simulation and the self-similarity property of Brownian motion.

Let

$$Y = f_n(B(1), B(2), ..., B(n))$$

be the payoff you get when exercising a path-dependent stock option for standard Brownian motion $B(t)$ and some given function f_n; our goal is to estimate $E[Y]$.

The process $X(t) = \exp\{at + bB(t)\}$ is called geometric Brownian motion with drift, and it is commonly used in finance as a model for a stock's price for the purpose of estimating the value of stock options. One example of a path-dependent option is the *lookback option*, with payoff function

$$Y = \max_{t=1,2,...,n} (\exp\{at + bB(t)\} - k)^+.$$

Another example, the *knockout option*, is automatically canceled if some condition is satisfied. For example, you may have the option to purchase a share of stock during period n for the price k, provided the price has never gone above a during periods one through n. This gives a payoff function

$$Y = (\exp\{an + bB(n)\} - k)^+ \times I\left\{\max_{t=1,2,...,n} \exp\{at + bB(t)\} < a\right\},$$

which is also path-dependent.

The usual method for simulation is to generate $Y_1, Y_2, ..., Y_n$ iid $\sim Y$, and use the estimator $\widehat{Y} = \sum_{i=1}^n Y_i$ having $\mathrm{Var}(\widehat{Y}) = \frac{1}{n}\mathrm{Var}(Y)$. The *control variates approach*, on the other hand, is to find another variable X with $E[X] = 0$ and $r = \mathrm{corr}(X, Y) \neq 0$ and use the estimator $\widehat{Y}' = \sum_{i=1}^n (Y_i - mX_i)$, where $m = r\sqrt{\mathrm{Var}(Y)/\mathrm{Var}(X)}$ is the slope of the regression line for predicting Y from X. The quantity m is typically estimated from a short preliminary simulation. Because

$$\mathrm{Var}(\widehat{Y}') = (1 - r^2)\frac{1}{n}\mathrm{Var}(Y) < \mathrm{Var}(\widehat{Y}),$$

we get a reduction in variance and less error for the same length simulation run.

Consider the simple example where

$$Y = \max\{B(1), B(2), ..., B(100)\},$$

and we want to compute $E[Y]$. For each replication, simulate Y, then

1. Compute $X' = \max\{B(10), B(20), ..., B(100)\}$.

2. Compute
$$X_0 = \sqrt{10}\max\{B(1), B(2), ..., B(10)\}$$
$$X_1 = \sqrt{10}(\max\{B(11), B(12), ..., B(20)\} - B(10))$$
$$\vdots$$
$$X_9 = \sqrt{10}(\max\{B(91), B(92), ..., B(100)\} - B(90)),$$
and note that self similarity of Brownian motion means that the X_i are iid $\sim X'$.

3. Use the control variate $X = X' - \frac{1}{10}\sum_{i=0}^{9} X_i$.

Your estimate for that replication is $Y - mX$.

Because $E[X] = 0$ and X and Y are expected to be highly positively correlated, we should get a low variance estimator.

7.4 Embedding Variables in Brownian Motion

Using the fact that both Brownian motion $B(t)$ and $(B(t))^2 - t$ are martingales (we ask you to prove that $(B(t))^2 - t$ is a martingale in the exercises at the end of the chapter) with continuous paths, the following stopping theorem can be proven.

Proposition 7.11 *With* $a < 0 < b$ *and* $T = \inf\{t \geq 0 : B(t) = a$ *or* $B(t) = b\}$, *then* $E[B(T)] = 0$ *and* $E[(B(T))^2] = E[T]$.

Proof Because $B(n2^{-m})$ for $n = 0, 1, ...$ is a martingale (we ask you to prove this in the exercises at the end of the chapter) and $E[|B(2^{-m})|] < \infty$, we see that for finite stopping times Condition 3 of Proposition 3.14 (the martingale stopping theorem) holds. If we use the stopping time $T_m = 2^{-m}\lfloor 2^m T + 1\rfloor$, this then gives us the first equality of

$$0 = E\left[B(\min(t, T_m))\right] \to E\left[B(\min(t, T))\right] \to E\left[B(T)\right],$$

where the first arrow is as $m \to \infty$ and follows from the dominated convergence theorem (using continuity of $B(t)$ and $T_m \to T$ to get $B(\min(t, T_m)) \to B(\min(t, T))$ and using the bound $|B(\min(t, T_m))| \leq \sup_{0 \leq s \leq t}|B(s)|$; this bound has finite mean by Proposition 7.9, and the second arrow is as $t \to \infty$ and follows again from the dominated convergence theorem (using continuity of $B(t)$ and $\min(t, T) \to T$ to get $B(\min(t, T)) \to B(T)$ and also using the bound $|B(\min(t, T))| \leq b - a$), and hence the first part of the result. The argument is similar for the second claim of the proposition, by starting with the discrete time martingale $(B(n2^{-m}))^2 - n2^{-m}, n = 0, 1, ... $
∎

Proposition 7.12 *With the definitions from the previous proposition,* $P(B(T) = a) = b/(b-a)$ *and* $E[T] = -ab$.

Proof By the previous proposition,

$$0 = E[B(T)] = aP(B(T) = a) + b(1 - P(B(T) = a))$$

and

$$E[T] = E[(B(T))^2] = a^2 P(B(T) = a) + b^2(1 - P(B(T) = a)),$$

which when simplified and combined give the proposition. ∎

Proposition 7.13 *Given a random variable X having $E[X] = 0$ and $\mathrm{Var}(X) = \sigma^2$, there exists a stopping time T for Brownian motion such that $B(T) =_d X$ and $E[T] = \sigma^2$.*

You might initially think of using the obvious stopping time $T = \inf\{t \geq 0 : B(t) = X\}$, but it turns out this gives $E[T] = \infty$. Here is a better approach.

Proof We give a proof for the case where X is a continuous random variable having density function f, and it can be shown that the general case follows using a similar argument.

Let Y, Z be random variables having joint density function

$$g(y, z) = (z - y)f(z)f(y)/E[X^+], \text{ for } y < 0 < z.$$

This function is a density because

$$\int_{-\infty}^{0} \int_{0}^{\infty} g(y, z) dz dy = \int_{-\infty}^{0} \int_{0}^{\infty} (z - y)f(z)f(y)/E[X^+] dz dy$$

$$= \int_{-\infty}^{0} f(y) dy \int_{0}^{\infty} zf(z) dz/E[X^+]$$

$$- \int_{0}^{\infty} f(z) dz \int_{-\infty}^{0} yf(y) dy/E[X^+]$$

$$= P(X < 0) + P(X > 0)E[X^-]/E[X^+]$$

$$= 1,$$

where we use $E[X^-] = E[X^+]$ in the last line.

Then let $T = \inf\{t \geq 0 : B(t) = Y \text{ or } B(t) = Z\}$. We then obtain $B(T) =_d X$ by first letting $x < 0$ and using the previous proposition in the second line here:

$$P(B(T) \leq x) = \int_0^\infty \int_{-\infty}^0 P(B(T) \leq x \mid Y = y, Z = z)g(y, z)dydz$$

$$= \int_0^\infty \int_{-\infty}^x \frac{z}{z-y}g(y, z)dydz$$

$$= \int_0^\infty \frac{zf(z)}{E[X^+]} \int_{-\infty}^x f(y)dydz$$

$$= P(X \leq x) \int_0^\infty \frac{zf(z)}{E[X^+]}dz$$

$$= P(X \leq x),$$

note that a similar argument works for the case where $x > 0$.

To obtain $E[T] = \sigma^2$, note that the previous proposition gives

$$E[T] = E[-YZ]$$

$$= \int_0^\infty \int_{-\infty}^0 -yzg(y, z)dydz$$

$$= \int_0^\infty \int_{-\infty}^0 \frac{yz(y-z)f(z)f(y)}{E[X^+]}dydz$$

$$= \int_{-\infty}^0 x^2 f(x)dx + \int_0^\infty x^2 f(x)dx$$

$$= \sigma^2.$$

∎

Remark 7.14 It turns out that the first part of the previous result works with any martingale $M(t)$ having continuous paths. It can be shown, with $a < 0 < b$ and $T = \inf\{t \geq 0 : M(t) = a \text{ or } M(t) = b\}$, that $E[M(T) = 0]$ and thus we can construct another stopping time T as in the previous proposition to get $M(T) =_d X$. We do not, however, necessarily get $E[T] = \sigma^2$.

7.5 Central Limit Theorem

We are now ready to state and prove a generalization of the central limit theorem. Because a sequence of iid random variables is stationary and ergodic, the central limit then follows from the following proposition.

Proposition 7.15 *Suppose $X_1, X_2 \ldots$ is a stationary and ergodic sequence of random variables with $\mathcal{F}_n = \sigma(X_1, ..., X_n)$ and such that $E[X_i|\mathcal{F}_{i-1}] = 0$*

and $E[X_i^2|\mathcal{F}_{i-1}] = 1.$ *With* $S_n = \sum_{i=1}^{n} X_i,$ *then we have* $S_n/\sqrt{n} \to_d$ $N(0,1)$ *as* $n \to \infty.$

Proof By the previous proposition and the strong Markov property, there must exist stopping times $T_1, T_2, \ldots$ where the $D_i = T_{i+1} - T_i$ are stationary and ergodic, and where $S_n = B(T_n)$ for Brownian motion $B(t)$. The ergodic theorem says $T_m/m \to 1$ a.s., so that given $\epsilon > 0$ we have

$$N_\epsilon = \min\{n : \forall m > n,\ m(1-\epsilon) < T_m < m(1+\epsilon)\} < \infty,$$

and so

$$
\begin{aligned}
P(S_n/\sqrt{n} \le x) &= P(B(T_n)/\sqrt{n} \le x) \\
&\le P(B(T_n)/\sqrt{n} \le x, N_\epsilon \le n) + P(N_\epsilon > n) \\
&\le P\left(\inf_{|\delta|<\epsilon} B(n(1+\delta))/\sqrt{n} \le x\right) + P(N_\epsilon > n) \\
&= P\left(\inf_{|\delta|<\epsilon} B(1+\delta) \le x\right) + P(N_\epsilon > n) \\
&\to P(B(1) \le x)
\end{aligned}
$$

because $\epsilon \to 0$ and $n \to \infty$ and using the fact that $B(t)$ is continuous in the last line. Because the same argument can be applied to the sequence $-X_1, -X_2, \ldots$, we obtain the corresponding lower bound and thus the conclusion of the proposition. ∎

7.6 Exercises

1. Show that if $X(t), t \ge 0$ is a continuous time martingale then $X(t_i), i \ge 0$ is a discrete time martingale whenever $t_1 \le t_2 \le \cdots < \infty$ are increasing stopping times.

2. If $B(t)$ is standard Brownian motion, show for any $a > 0$ that $B(at)/\sqrt{a}, t \ge 0$ is a continuous time martingale with stationary independent increments and $B(at)/\sqrt{a} =_d B(t)$.

3. If $B(t)$ is standard Brownian motion, compute $\text{Cov}(B(t), B(s))$.

4. If $B(t)$ is standard Brownian motion, which of the following is a continuous time martingale with stationary independent increments? (a) $\sqrt{t}B(1)$, (b) $B(3t) - B(2t)$, or (c) $-B(2t)/\sqrt{2}$.

5. If $B(t)$ is standard Brownian motion, show that $(B(t))^2 - t$ and $B^3(t) - 3tB(t)$ are continuous time martingales.

6. If $B(t)$ is standard Brownian motion and $T = \inf\{t > 0 : B(t) < 0\}$, compute $E[T]$.

7. Is $(N(t))^2 - \lambda N(t)$ a martingale when $N(t)$ is a Poisson process with rate λ?

8. Prove the strong Markov property for Brownian motion as follows: (a) First prove for discrete stopping times T using the same argument as the strong Markov property for Markov chains. (b) Extend this to arbitrary stopping times $T < \infty$ using the dominated convergence theorem and the sequence of stopping times $T_n = (\lfloor 2^n T \rfloor + 1)/2^n$. (c) Apply the extension theorem to show that Brownian motion after a stopping time is the same as Brownian motion.

References

The material from Sections 2.4, 2.6, 2.5, and 2.7 comes from references 1, 3, 5, and 6, respectively. References 2 and 4 give more exhaustive measure-theoretic treatments of probability theory, and reference 7 gives an elementary introduction without measure theory.

1. Barbour, A. D., Holst, L., and Janson, S., *Poisson Approximation,* Oxford University Press, New York, 1992.

2. Billingsley, P., *Probability and Measure,* 3rd ed., John Wiley & Sons, Inc., New York, 1995.

3. Chen, L., and Shao, Q., "Stein's Method for Normal Approximation," in *An Introduction to Stein's Method,* ed. by Barbour, A. D., and Chen, L., World Scientific, Hackensack, NJ, 2005.

4. Durrett, R., *Probability: Theory and Examples,* 2nd ed., Duxbury Press, Belmont, CA, 1996.

5. Peköz, E., "Stein's Method for Geometric Approximation," *Journal of Applied Probability,* Vol. 33 (1996), 707–713.

6. Peköz, E., and Röllin, A., "New Rates for Exponential Approximation and the Theorems of Rényi and Yaglom," *Annals of Probability,* Vol. 39 (2011), No. 2, 587608.

7. Ross, S. M., *A First Course in Probability,* 10th ed., Prentice Hall, Upper Saddle River, NJ, 2019.

Index

Printed in the United States
by Baker & Taylor Publisher Services